# NATURA

## Biology for Bilingual Classes

edited by

Rudolf Marx

## Ecology

Ernst Klett Verlag
Stuttgart · Leipzig

1. Auflage

1 8 7 6 5 | 2024 2023 2022

Alle Drucke dieser Auflage sind unverändert und können im Unterricht nebeneinander verwendet werden. Die letzte Zahl bezeichnet das Jahr des Druckes.


Internetadresse: www.klett.de

**Bearbeitet von**
Rudolf Marx, Head of Key Stage 3 Science, The Netherhall School and Sixth Form Centre, Cambridge/UK

**unter Mitarbeit von**
Dr. Irmtraud Beyer, Dreieich; Dr. Horst Bickel, Düsseldorf; Roman Claus, Rees; Roland Frank, Stuttgart; Prof. Dr. Harald Gropengießer, Hannover; Gert Haala, Wesel, Prof. Dr. Siegfried Kluge; Neumark; Bernhard Knauer, Göttingen; Dr. Inge Kronberg, Hohenwestedt; Hans-Peter Krull, Kaarst; Hans-Dieter Lichtner, Stadthagen; Uschi Loth, Burbach; Dr. Horst Schneeweiß, Hamburg; Dr. Jürgen Schweizer, Stuttgart; Ulrich Sommermann, Münchberg; Gerhard Ströhla, Münchberg; Dr. Wolfgang Tischer, Sarstedt; Günther Wichert, Dinslaken

**Redaktion:** Dr. Peter R. Menke

**Mediengestaltung:**
Ingrid Walter

**Fachwissenschaftliche und sprachliche Beraterin:**
Dr. R. Theresa Jones, Wolfenbüttel

**Lautschrift:**
Peter Bereza, Aachen

**Layoutkonzeption und Gestaltung**
Prof. Jürgen Wirth; Visuelle Kommunikation, Dreieich
unter Mitarbeit von Matthias Balonier, Evelyn Junqueira, Nora Wirth

**Umschlaggestaltung**
Jens-Peter Becker; normal Industriedesign, Schwäbisch Gmünd; unter Verwendung von folgendem Foto: Bildagentur Mauritius (age fotostock), Mittenwald

**Reproduktion**
Meyle + Müller, Medien-Management, Pforzheim

**Druck**
Digitaldruck Tebben GmbH, Biessenhofen

Printed in Germany

ISBN 978-3-12-045341-3

## Warning signs and experiments in school

Natural sciences like biology are not imaginable without experiments. Natura Oberstufe contains also some experiments.

However, experimenting with chemicals is never completely safe. Therefore, it is important to discuss potential sources of danger with your teacher before starting the respective experiment. Especially in the laboratory obvious behavioural rules always have to be observed. The precautions taken are dependent on the danger potential of the used substance.

For that reason the chemicals are marked with a warning sign in the experimental manual. The warning signs are also present on the label of the chemical's container.

They mean:

**C** = *corrosive*
Living tissue and materials that come in contact with this substance will be destroyed at the affected site.

**F** = *flammable*
Substances that are easily ignited when exposed to an ignition source.

**Xi** = *irritating* (X stands for St. Andrew's cross)
Substances that can be irritating to skin, eyes or respiratory organs.

**Xn** = *noxious*
Substances that can cause health problems when inhaled, swallowed or contacted the skin.

## What is in this book?

Ecology is the scientific study of the interactions between organisms and their environment on various organisational levels.

This book covers all relevant aspects of ecology for biology in the "Oberstufe". It provides you with a thorough and well-structured introduction into this highly complex and important subfield of biology. The book first focuses on the interactions between individual organisms and their environment; then the view is gradually widened to include the ecology of populations, the concept of ecosystems and eventually the big picture of the biosphere as whole with humankind as one of its most influential factors. Through this bottom-up approach you will start to grasp the complexity of ecology and start to think in ecological categories for themselves. By hypothesising and testing ideas, by applying and connecting knowledge, theories and models and by understanding the concept of interdependence, you will not only enter the world of scientific thinking, but master a skill that will be crucial for solving the problems of the 21st century.

In order to make this process easier, this book contains many pages which not only provide information, but also motivate students to discuss, explore and question scientific topics.

**Info pages**
These provide basic information about a topic. Numerous figures illustrate complex issues. The short tasks test whether you have understood the text, or not.

**CD**
The answers to the tasks are on a CD, which also contains word lists (with pronunciation) to help you to learn in a foreign language.

**»info box«**

In info boxes you find interesting examples, exceptions, methods, etc.

**Practicals pages**
Here experiments are described that you can carry out yourself. It is important to work precisely and to make accurate reports in order to obtain the best results.

**Encyclopedia pages**
These pages provide additional information which place the topics you have learned about in a broader context and in that way give an overview. Topics are included that are not mentioned directly in the teaching curriculum. These pages are meant to encourage further reading.

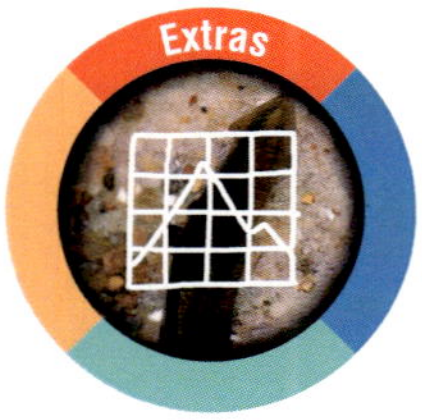


**Extras pages**
Extensive information about a specific topic and corresponding tasks are provided on these pages.

**Issues pages**
Biological problems often go beyond the subject of biology.
These pages provide interdisciplinary material and impulses to work independently. They also refer to topics in everyday life.

**Basic concepts**
Biology is a complex science and its different disciplines are connected by many, partly abstract, relations. This new type of page shows interconnections between very different topics. Basic concept pages clarify principles and enable you to order and structure known facts. The tasks encourage pupils to discuss and to think about examples again.

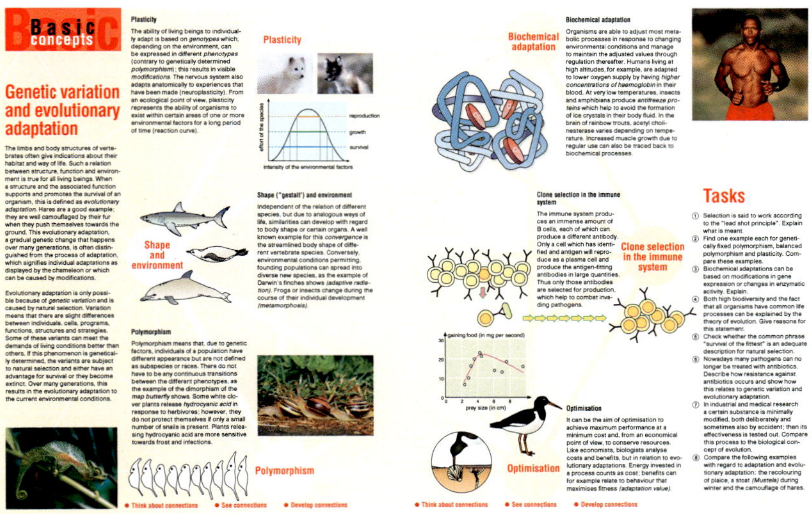

# Foreword

Welcome to the second book in this series of the bilingual programme for the "Oberstufe". Bilingual classes are a challenge for both pupils and teachers. Most scientific articles are written in English and the language provides a link between researchers around the world and the knowledge they have. One can say that English has become the "lingua franca" of science. Therefore, it is essential to be comfortable with English and to learn commonly used scientific terms. This book wants to take away the scepticism, and maybe even the fear that teaching and learning in a foreign language might cause. The level of English used in the book corresponds to that expected in the "Oberstufe". Additionally, pupils should find the word lists for each topic (on the CD) helpful. The layout of this book corresponds to chapter 6 of the German textbook, "NATURA, Biologie für Gymnasien, Oberstufe", that can always be used for reference purposes.

This volume introduces one of the classical disciplines of biology: ecology. Ecology is the scientific study of the interactions between organisms and their environment.
Although a lot of current biological research is happening in the fields of genetics and biochemistry, ecology is by no means out-dated. Any of the global issues highlighted in the news on an almost daily basis can be explained through ecology: be it loss of tropical rainforests and their biodiversity, be it global warming or be it the fact that the world population is growing at an ever increasing rate. Like any other living being, humans impact on their environment, and the environment in turn has an impact on all the living organisms in it. The major difference between us and any other living species, however, is that we are changing our environment permanently, rapidly and in a potentially unsustainable way.
Any attempt to revert damage already done and to live in a sustainable way, both as an individual and as a society, requires a thorough understanding of the complex interaction of all living things and their environment; this is what ecology and its subdisciplines describe and analyse.

The role of humans as the major cause of extinction becomes unambiguous. Palaeontological studies, for example, show that the Hawaiian Islands probably had more than 100 species of native birds when Polynesian colonists arrived about 300 A. D. However, that number had been reduced by half by the time European colonists reached the American Island Hawaii.
Without a thorough understanding of how species work, management efforts might have negative effects on the species concerned. Therefore, enormous amounts of data about the basic biology of the species are needed. Organismal biology and conservation work hand in hand.
Conservation efforts are truly complicated by economic and political issues. Furthermore, protective strategies being appropriate for developed nations may be impractical in developing countries. Only programs that integrate the needs of wildlife and of humans have the best chances for a long-time success.

A true understanding of ecology and the concept of interdependence however also makes us painfully aware of the limitations of this powerful discipline: because life on Earth is so incredibly complex, diverse and interdependent, it is virtually impossible to exactly predict the outcomes of anything we do. We can make very well founded predictions, produce sophisticated mathematical models and scenarios, but in the end only reality will prove our theories right or wrong.

The current debate about the reasons for global warming is a good example: are our increased carbon emissions causing global warming and climate change? There is a strong divide among learned scientists on this question. But: can we risk following the advice of those who say that carbon emissions do not cause global warming? What if it turns out in the end that they were wrong? It might be too late.
Nevertheless – despite or just because of this dilemma: the best possible ecological understanding is more important than ever in order to be able to face the global challenges of the 21st century.

Communication will play a crucial part in this major effort: scientists need to share and discuss their findings, hypotheses and theories; politicians need to negotiate agreements, economists need to react and individual members of society need the scientific literacy to understand and to judge for themselves. The common language will be English – you had better be prepared.

More books presenting essential knowledge like "Genetics and Immune System" or "Neurobiology" will be published in the near future in order to expand the possibility of studying one of the most important natural sciences.

Rudolf H. Marx
Susanna Bächle
Doris Bächle-Knauer
Dr. Peter R. Menke

July 2008

# Contents

## Ecology

# Ecology

The complexity of life on Earth calls for a variety of perspectives for the observation of biological phenomena. Cytology, physiology, and molecular biology deal with the structure and function of a living being and its organs.

All living beings however exist in a close relationship with their *environment* – a term describing everything that exists outside the living individual. These relationships are collected into the subject of *ecology*. Different points of view and perspectives can also be applied here.

The relationships between a single individual and its environment are the main topic of *autecology*. Autecological studies mainly require close links with the field of physiology.

*Conspecifics* and organisms of different species certainly form part of the environment of a living being. *Population ecology* focuses on this perspective.

The relationships between living beings of different species can be highly diverse: individuals of different species can promote the vitality of a living being, but they also might inhibit it. These *interactions* between different species are highly complex and are described by *synecology*, which itself is closely linked to population ecology.

autecology

population ecology

synecology

Populations of different species in turn exhibit a close relationship with their specific living space. They influence it, and the living space influences the populations living within it. Consequently, complex interactions occur between all cohabiting living beings (*biocenoses*) and their living space (*biotope*), the environment with all its determining physical and chemical factors. The *research on ecosystems* deals with this complex interplay.

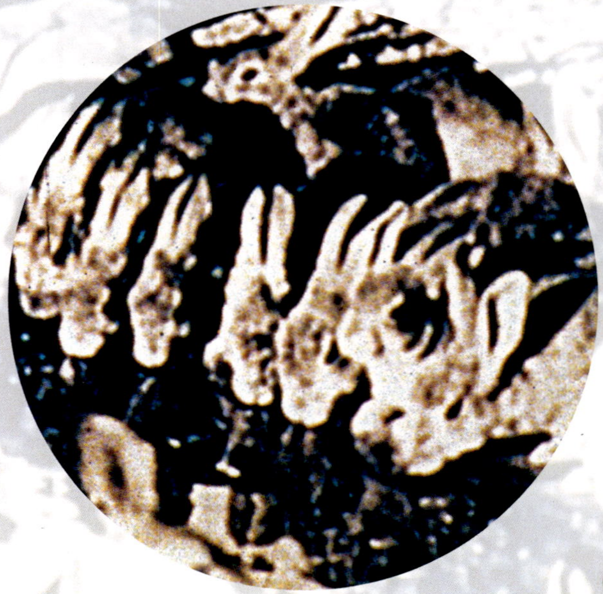

Synecology and population ecology also have close links to other fields of research in biology. Above all, aspects of behavioural biology are closely intertwined with ecological research (*ecoethology*). In turn, all these subdisciplines are finally comprehensible only when considered alongside the evolutionary history of populations. On the one hand different perspectives prove useful as they provide an overwiew. On the other hand, only a *synoptic* view of all branches of research allows a comprehensive understanding of ecological phenomena. For this reason the same phenomena are scrutinised from different points of view.

Human beings are also integrated into their environment and are dependent upon it in many ways. However, more than any other living being, humans actively interfere and change their environment; man-made natural catastrophes are only the most visible effects. More than that by any other living being, this interference by humans endangers the existence of other species. This interaction is not necessarily a characteristic of humans only but, because of its outstanding significance for us and for all other ecosystems, the scientific discipline of *human ecology* is separately devoted to the specific conditions of this relationship.

## The influence of temperature on living beings

**Biocenosis**
The community of all living beings in a biotope

**Biotope**
The environment: the sum of all abiotic environmental factors

**Habitat**
The specific environment favoured by a single species

One of the main objectives of ecology is to understand the interdependence between a living being and its environment. This explains why, for example, the source of a stream has a completely different *biocenosis* than a pond. The determining impacts are both *abiotic* and *biotic*. Physical and chemical factors such as temperature, pH value or salt content are abiotic factors; organisms living in the same living space (*biotope*), e.g. *prey* and *predators*, are equally important as biotic factors.

**1** Emperor penguins in Antarctica

With regard to temperature, two main *adaptations* are found as a matter of principle: *Homoeothermic* (warm-blooded) organisms, such as birds and mammals, *regulate* their body temperature within given limits at a physiological optimum between approximately 36 °C and 40 °C, more or less independent of the temperature of the surrounding. Insulation provided for example by structures like *plumage*, coats or *blubber* minimises heat loss. Control mechanisms regulate the body temperature of homoeothermic organisms through *homeostasis* by using changes in the *metabolic rate* and the rearrangement of blood vessels. A specific arrangement of blood vessels allows penguins and other polar animals to minimise heat loss and to walk permanently on ice without freezing to death.

*Poikilothermic* (cold-blooded) animals do not maintain their body temperature at constant levels. Body insulation structures are rarely found with these animals. Their body temperature largely corresponds to the outside temperature. These *conformers* show physiological adaptations such as heat-resistant proteins or enzymes that attain their metabolic optimum already at low temperatures. Usually, poikilothermic animals actively seek areas with more comfortable temperatures, be it escaping underground in hot deserts or lying in the sun in cold regions.

Living beings are adapted to their specific *habitat*, i.e. they show features that help them to survive in the conditions present. One important factor can be demonstrated by temperature. On Earth, living spaces of highly different temperatures exist — and all of them are populated by living beings. Penguins can bear up to -60 °C in Antarctica, whereas some microorganisms populate hot volcanic springs of up to 90 °C (fig. 2) and, under deep ocean conditions, they can even withstand temperatures of up to 120 °C. Temperature extremes are problematic as virtually all physiological processes are tied up to enzymatic reactions. The reaction rate-temperature rule (van't Hoff's rule) and the denaturing of proteins caused by heat mainly fix the physiological limits.

**2** Bacteria in hot springs

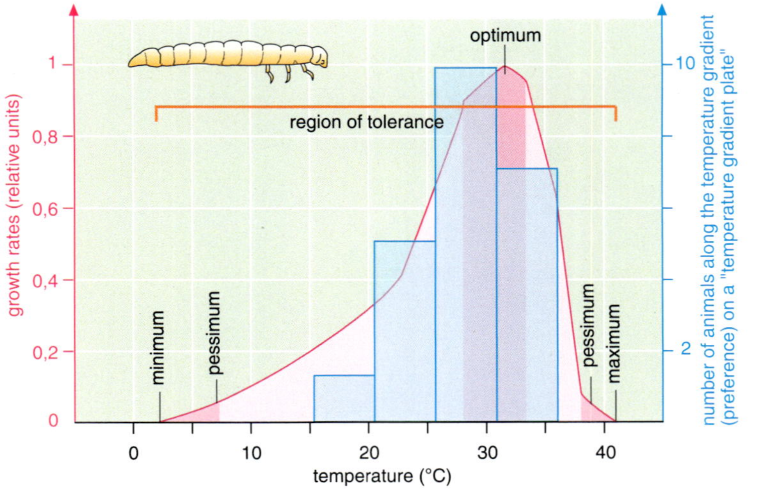


**1** Temperature tolerance and preference of the mealworm

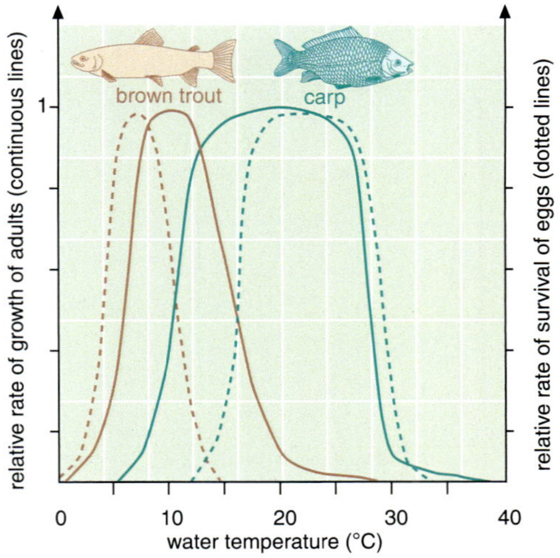


**2** Ecological potencies

**Ecological potency**
Extent of response of a species to an environmental factor (feature of a living being)

**Region of tolerance**
Environmental conditions under which a living being can exist

**Stenopotency**
(Gr. *sten* = small, Lat. *potentia* = potential) Limited tolerance with respect to an environmental factor

**Eurypotency**
(Gr. *eurus* = wide) High tolerance

## Ecological potency

Ecological potencies allow a quantitative description of the requirements of a species. They can be described for any *abiotic influence*. This can be demonstrated by examining the dependency on temperature.

*Brown trout* populate the headwaters of cold mountain streams. They spawn between November and December. *Carp* on the other hand prefer warm standing or only slowly moving bodies of water. Spawning time is between May and June. Species-specific requirements like these can also be tested in experiments: in *preference testing*, some individuals of a species are placed on a temperature gradient plate where they can choose their preferred resting place (fig.1). Admittedly, in practice this is more difficult with animals larger than fish; *tolerance tests* are employed (fig. 2) here, which can also be used with plants: over a longer period of time the living beings are exposed to a defined temperature, while their *vitality*, i. e. their ability to live, is measured by using mortality (death) or survival rates. Most graphs of a species' tolerance are bell-shaped with an upper (maximum) and a lower (minimum) temperature limit and a temperature optimum (fig. 1). The range between the minimum and the maximum value is called the *tolerance region*. This is important as environmental conditions (e. g. temperature) can sometimes vary considerably in certain areas. At the upper and lower limits, reproduction is usually not possible any more (*pessimum region*). Under such conditions, a species cannot survive in the long term. The *area of preference* is usually narrower than the tolerance region. Tolerance curves are determined for either individuals or for whole populations.

Genetic differences can cause variation between some individuals of a species with regard to their ecological potencies. The tolerance region of a whole population is therefore wider than that of a single individual. This tolerance of the whole population is of crucial significance for the adaptability of a population when environmental conditions change.

The ecological potency of the brown trout regarding temperature (rate of survival of eggs and growth of adults) is narrower than that of carp. A brown trout thus only has a limited tolerance, i. e. *stenopotency*, to temperature. Carp on the contrary show a comparatively high tolerance; regarding temperature carp are considered *eurypotent*.

### Task

① Compare the occurrence and the spawning times of both the brown trout and carp with the experimental data (fig. 2).

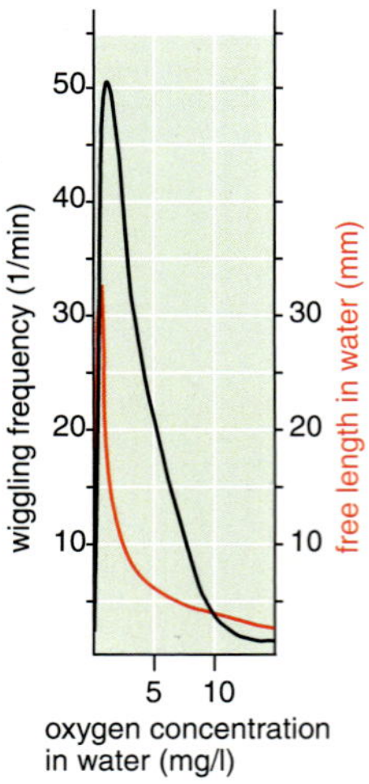


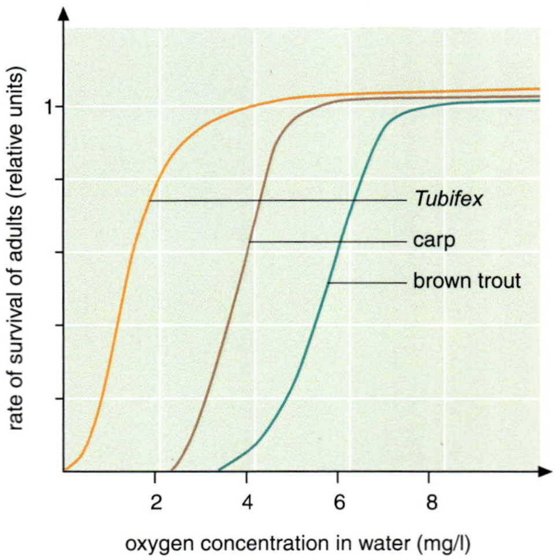


**1** Oxygen demand

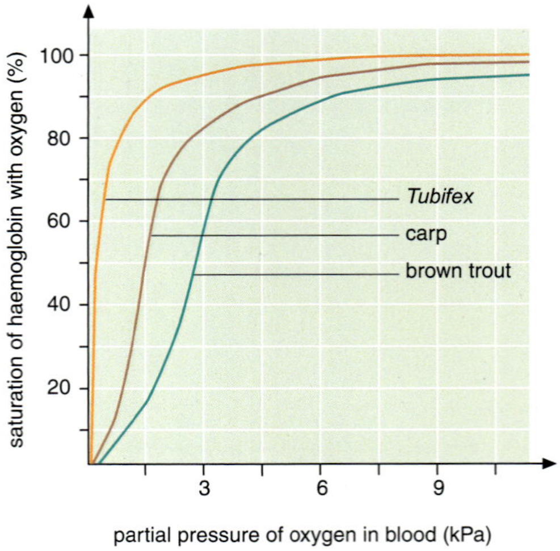


**2** Oxygen saturation of haemoglobin

## The effect of oxygen on animals

oxygen concentration per litre at 20 °C:

air (at sea level)
273 mg/l

air (5000 m altitude)
145 mg/l

fresh water
0 – 9 mg/l

sea water
0 – 5 mg/l

Most animals need *oxygen* for *respiration*. The percentage of oxygen in the air is the same almost everywhere (except at great heights). In water, however, it can vary to a great extent which is why oxygen is a *limiting factor* in water. Contrary to temperature, an insufficient oxygen supply is harmful to most animals (except for anaerobic microorganisms), whereas oxygen excess is not damaging (fig. 1).

The oxygen content of water is dependent on another environmental factor, i. e. temperature. As cold-blooded animals need more oxygen with increasing temperature (*reaction rate-temperature rule*; also called *van't Hoff's rule*), temperature becomes limiting in two ways: with increasing temperature oxygen supply decreases, while the demand for oxygen goes up at the same time.

In standing water the concentration of oxygen can vary significantly with depth. At the bottom of strongly polluted waters, hardly any oxygen is found as abundant bacteria use up any oxygen for their respiration. Only highly specialised organisms such as the sludge worm *Tubifex*, a close relative of the common earthworm, can survive there. It feeds on organic matter and bacteria, both of which are in large supply. Like some other *invertebrates* that live in sewage, their blood contains dissolved *haemoglobin*. This *skin-breather* has a surprising behavioural pattern: the lower the oxygen concentration, the further it emerges from the sediment and the more it wiggles its body (see margin) in order to increase the amount of fresh water passing over its skin. When the oxygen concentration becomes too low, *Tubifex* stops wiggling and switches to anaerobic respiration for its energy supply. This behaviour makes *Tubifex* well adapted to variable concentrations of oxygen — it is eurypotent.

Some homoeothermic land animals have similar adaptations: the blood of lamas living at high altitudes in the Andes contains a higher haemoglobin concentration than blood of related species living at lower altitudes. The oxygen affinity of the haemoglobin is similarly increased. Birds, which require enormous amounts of energy during flight, have developed strong efficient respiratory systems: air sacks that are connected to the lung enable them to take in more oxygen.

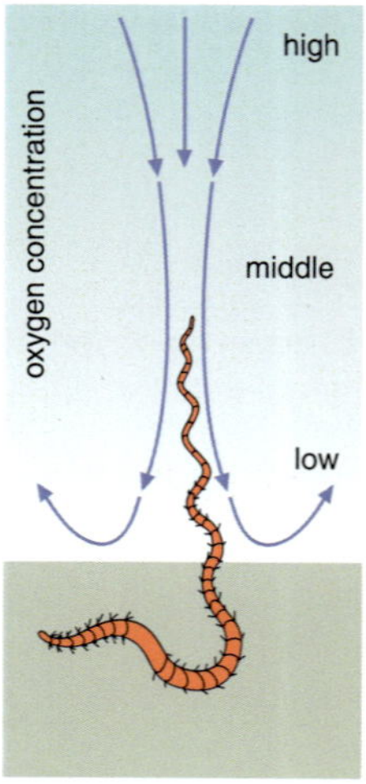


Tubifex

### Tasks

① Draw up a table to compare the adaptations of *Tubifex* at different oxygen concentrations in water.

② Give reasons for the high rate of survival of *Tubifex* at low oxygen concentrations in water (fig. 1). Use the data in figure 2.

③ Trout, which are predators, have a higher demand for oxygen than the peaceable carp (0.39 mg $O_2$/kg body mass for trout vs 0.12 mg $O_2$/kg body mass for carp). Establish a link to the data displayed in figures 1 and 2.

# Life with little water – the kangaroo rat

Water is essential for all living beings as physiological processes such as transport and biochemical reactions occur in aqueous solution. Thus, life in regions with very little available water requires special adaptations. The *kangaroo rat* (*Dipodomys merriami*) from Arizona (USA) is one example.

## Habitat and behaviour

This 24-cm-long animal lives on dry seeds that are rich in fat. Like many other *desert animals*, it moves about by jumping. This way of moving forward is an adaptation to open, unprotected habitats. By using comparatively little energy, the animal can move forward quickly and travel long distances, an advantage given that its food is widely scattered. Jumping would be less favourable in dense plant vegetation.

The average humidity of air in regions in which kangaroo rats occur can vary between 3 % and 50 % (average 25 %).

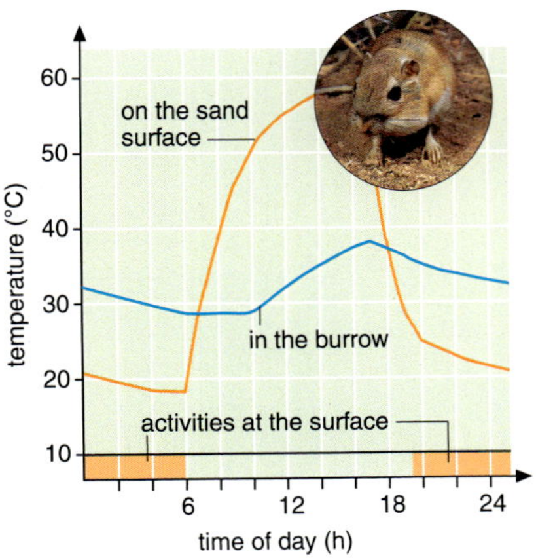


Rainfall is so rare that, most of the year, no water is available for drinking. During the day, the kangaroo rat stays in its burrow, which can be up to 1 m deep.

### Task

① Interpret the diagrams by using the text.

## Structure of the nose

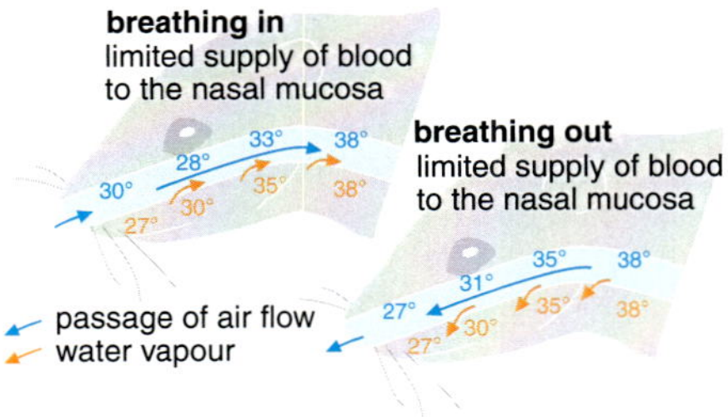


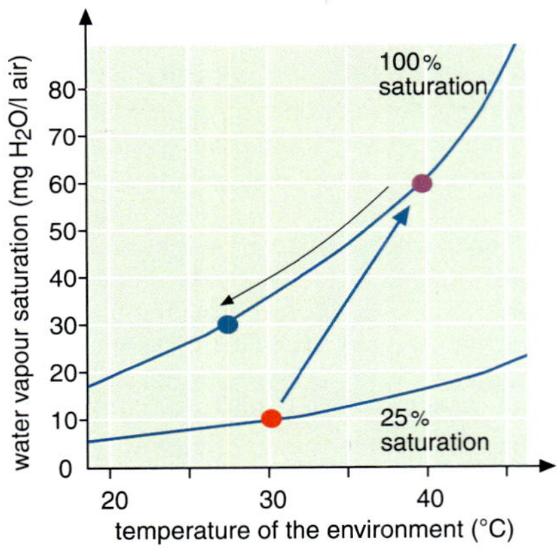


● exhaled air: relative humidity = 100%
● air saturated with water vapour in the lungs = 38%
● inhaled air: relative humidity = 25%

In comparison with other mammals of a similar size, the temperature in the *nasal cavity* of kangaroo rats is low. The airway, mainly the *nasal mucosa*, is poorly supplied with blood. Contrary to other animals, the nasal mucosa can absorb large amounts of water. When dry it is *hygroscopic* (water-attracting).

### Tasks

② Describe and explain the changes of the air that a kangaroo rat inhales and exhales. Take into account that evaporation of water has a cooling effect.

③ The nasal mucosa of the brown rat has a better blood supply. The cooling effect of the inhaled air is quickly compensated by the higher blood temperature.
Establish a hypothesis regarding the effects of these differences compared with the kangaroo rat.

④ Describe the consequences of this difference with regard to the loss of water in the process of exhaling.

## Metabolism and excretion

During most of the year, the only food for the kangaroo rat are seeds. These are rich in fat, but have a low water content. Respiration however is an alternative water supply.

Gain of water through respiration from 100 g of food:

| | |
|---|---|
| carbohydrate: | 60 g |
| fat: | 109 g |
| protein: | 44 g |

### Tasks

⑤ Compare the water gains and water losses of both the kangaroo rat and the brown rat and explain the differences by using all materials provided.

⑥ Make a list of all the features mentioned in the text showing how the kangaroo rat is adapted to life in a dry desert.

| | | Kangaroo rat | | Brown rat | |
|---|---|---|---|---|---|
| Composition of food | | Protein | 12 % | Protein | 22 % |
| | | Digestible carbohydrates | 8 % | Digestible carbohydrates | 50 % |
| | | Fat | 40 % | Fat | 4 % |
| Water balance for every 100 g of food | intake | Food | 6 g | Drinking | 25 g |
| | | Water from oxidation | 54 g | Food | 12 g |
| | | | | Water from oxidation | 46 g |
| | | | **60 g** | | **83 g** |
| | loss | Breathing and evaporation | 44 g | Breathing and evaporation | 50 g |
| | | Urine | 13.5 g | Urine | 25 g |
| | | Faeces | 2.5 g | Faeces | 8 g |
| | | | **60 g** | | **83 g** |

## More abiotic factors

### Light

**1. Light as a source of energy**

Characteristic structures make leaves adapted to different light *exposure* as either *shade* or *sun leaves*. Animals can also use light as an energy source: dark body surfaces in particular absorb light and transform it into heat on the skin.

**2. Light as a signal generator**

For both animals and plants, the duration of light exposure can be a signal to *synchronise* developmental processes or other activities with the time of day or the seasons.

Example: *germination of seeds*
For some plants such as poppy or the wild rose (so-called *light germinators*), a short light stimulus of a few minutes is enough to initiate germination. For *dark germinators* such as pumpkins, on the other hand, light inhibits the germination process. This ensures that seeds only germinate in the conditions to which they are adapted.

In many cases, animals orient themselves by using visual stimuli and therefore need light to make sense of their surroundings. However, behaviour is often determined by potential predators; nocturnal animals avoid daylight and so decrease the risk being hunted by their predators. The *annual cycle* with its changes of light conditions governs animal behaviour with regard to reproduction and *migration* (birds are a good example).

### Salt content

The salt content of the oceans is mainly determined by the concentrations of $Na^+$ and $Cl^-$ ions and has an *osmotic effect* on living cells.

*Homoiosmotic* organisms are *osmoregulators*; their internal ion concentration remains almost constant. (Figure 1: The lowering of the freezing temperature ($\Delta T$) is a measure of the concentration of dissolved salts): osteichthians (bony fish) living in oceans constantly regulate the ion concentration in their blood to a value that is *hypotonic* in comparison with the surrounding sea water. They therefore constantly lose water through *osmosis*. Excess ions are actively excreted through their gills. For bony fish in fresh water, on the contrary, the kidneys are more important. Since they actively take in ions through their gills, their inner milieu is *hypertonic*; water is constantly entering their bodies via osmosis. The kidneys produce large amounts of hypotonic urine and so maintain the ion concentration. Many coastal plants and sea birds excrete excess salt through salt glands.

When starfish end up in river estuaries, they start to swell. Like many other oceanic invertebrates, they do not have a developed system for the regulation of ion concentration. The concentration inside their bodies changes to reflect the conditions of the surrounding water. These *osmoconformers* are poikilosmotic and usually depend on an almost constant ion concentrations in their surroundings (fig. 2).

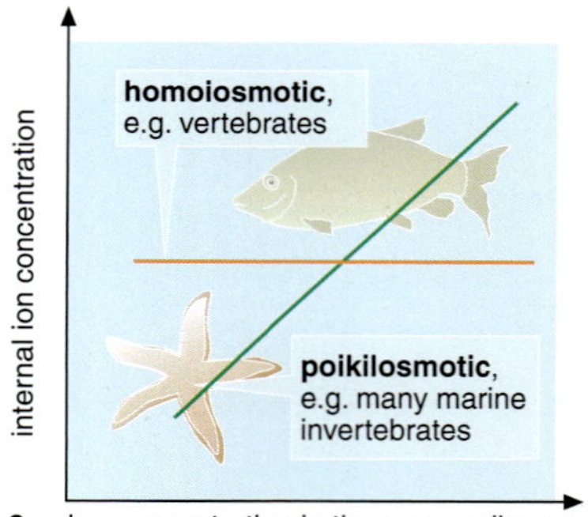


**2**

### Mineral concentration and pH value

*Minerals* that are essential for plants are partly bonded to soil particles and occur in equilibrium with the ions dissolved in the water, resulting in an almost constant concentration over long periods of time. The availability of mineral ions strongly depends on the pH value. If the soil particles cannot bond many of the important ions, the soil is poor in minerals. At a low pH value, toxic ions such as $Al^{3+}$ can become soluble and poison some plants. This acidification is one of the reasons for the damage to many forests and trees by *acid rain*.

Some plants prefer a low pH value, (e.g. sphagnum, a kind of moss: pH 3 – 4). Other plants can be found on alkaline soil (e.g. coltsfoot: -pH 8). The occurrence of certain stenopotent plants often gives an indication about the composition of the soil; these plants are referred to as *indicator plants*.

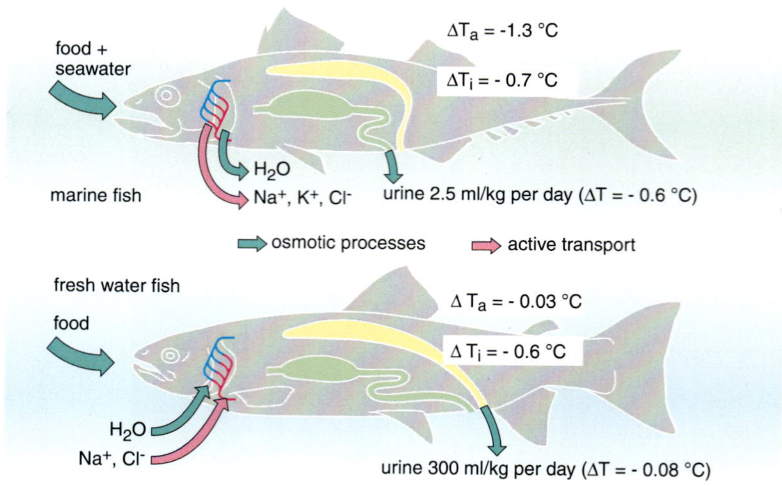


**1** Osmoregulation in salt and fresh water fish

| Subspecies | Body mass in kg* | Body length in cm* |
|---|---|---|
| Siberian tiger | 250 kg | 300 cm |
| Bengal tiger | 235 kg | 290 cm |
| South China tiger | 150 kg | 250 cm |
| Indochinese tiger | 170 kg | 270 cm |
| Sumatran tiger | 120 kg | 230 cm |
| Balinese tiger (†) | 95 kg | 225 cm |

*(male adult animal)

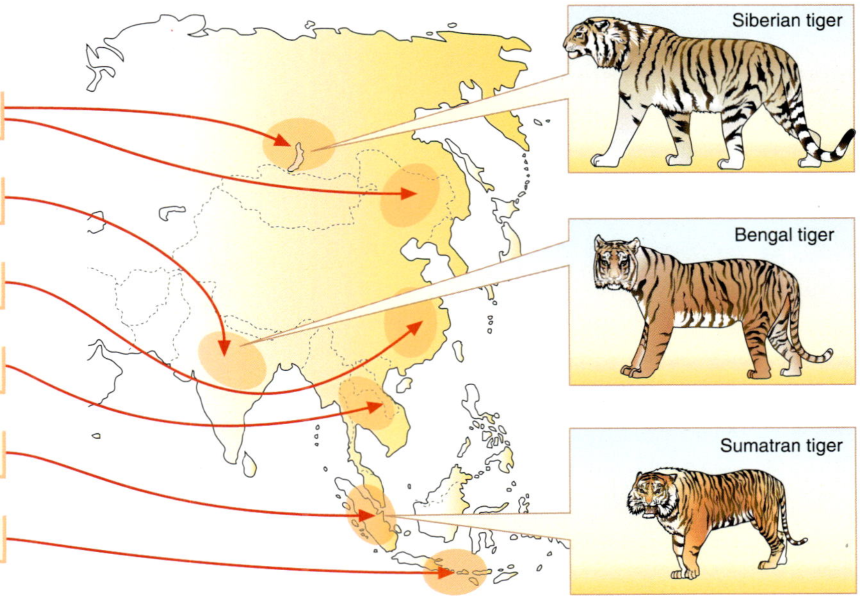


**1** Spread and features of the various geographical subspecies of the tiger

## The rules of zoogeography

jack rabbit

hare

alpine hare

desert fox

The black-tailed jack rabbit (*Lepus californicus*), in America also simply referred to as the jack rabbit, has strikingly large ears. The jack rabbit is found in the Great Plains and deserts in the southwest of the United States and in Mexico. J. A. Allen (1838 – 1921) was able to prove that large ears help the jack rabbit with its *thermoregulation*. Their ears are well supplied with blood; as soon as the outside temperature is slightly below their body temperature, they prick up their ears and can use their *dilated* blood vessels to give off heat.

Closely related species such as the mountain hare (*Lepus timidus*), on the contrary, have comparatively small ears as they live in cold climates; they also have shorter legs. Based on a smaller overall surface area, they can minimise heat loss. These zoogeographical observations have been generalised in *Allen's rule*: protruding body parts such as the ears or tails of warm-blooded species living in cold climates are usually smaller than the corresponding body parts of related species inhabiting warmer climates.

It is important to beware of over-simplified explanations, because large ears are of advantage for *crepuscular* animals such as the *fennek* (desert fox) as they help to locate both predators and potential prey. Many desert animals such as the desert fox live in burrows during the day where thermoregulation is possibly only of minor importance.

Apart from limbs and body parts, overall body size determines the relative body surface. Small warm-blooded animals have a comparatively larger relative body surface area than bigger animals of similar body structure and shape and they lose heat more quickly, a feature that is a disadvantage in colder climates. This observation is generalised in *Bergmann's rule*: the size of closely related warm-blooded species or subspecies increases going from warmer climates to the arctic regions. The subspecies of the tiger are a good example: they inhabit markedly different climatic zones and living spaces, starting from the north Asian tundra to the tropical jungles in south Asia.

However, in addition to *the balance of thermoregulation* other factors also influence body size. It is easier for smaller animals to move in thick jungles in the south than it is for larger animals. Larger animals have more extensive fat reserves that help them survive the long and snowy winters in the bare regions of north Asia, where food might be in short supply and they sometimes have to survive without food for considerable lengths of time.

# The concept of ecological niches

diatoms

*River nerites* (a type of snail) are about 1 cm in length and used to be quite common in the rivers of central Europe. They make few demands on their environment. Both temperature and pH value can vary greatly; they are sufficiently euypotent with regard to salt concentration that they even used to occur in the brakish water of large rivers. They feed on *diatoms* that they scrape off from the substrate with their *radula*, a rough tongue. Many species of hard-shelled diatoms are widespread so that these river nerites should never be short of food. It was hard to understand therefore, why river nerites became more and more rare in rivers and have not at all been found in the Rhine since 1994.

**Ecological niche**
Total of all requirements of a species regarding its biotic and abiotic factors in its environment

The biologist DIETRICH NEUMANN investigated this question further. He put river nerites into an aquarium filled with water only and gave them diatoms as food.
Whereas diatoms occupied the water and the glass walls, the river nerites quickly lost weight, although they had diatoms in their stomachs. However, the shells of these diatoms were undamaged and therefore could not be digested by the river nerites. It was only when NEUMANN added rough pebbles to the aquarium that the shells of the diatoms broke when the river nerites scraped them off the ground. Over the past few decades, the rocks in rivers have increasingly become overgrown by algae because of human influences. On this slippery surface, the river nerites cannot use the hard-shelled diatoms as food.

The total of all requirements made by the river nerite on its environment is considered its *ecological niche*. This concept not only denotes a space — the actual living space of the river nerite, i.e. the river and rocky riverbed, is considered its *habitat*. The habitat is of course part of the ecological niche, but in addition to this, it also comprises requirements regarding oxygen concentration, temperature, pH, etc.

The concept of the ecological niche can be differentiated into various parts. In addition to diatoms, the river nerites also need a rough substrate as a feeding requirement (*food niche*). Other parts of the niche that should be considered are e. g. demands for nesting spaces (*breeding niche*) or hiding places (*hiding niche*).

The example of river nerites shows that environmental conditions can have variable significance for a living being. The speed of a river current is less important for this species as they can attach themselves firmly to the ground and have a streamlined shape: the river nerites are eurypotent with regard to this factor. One requirement, in this case the rough ground surface, can however be crucial for the survival of a species.
The "*law of the minimum*" applied to environmental factors describes this relationship: the environmental factor that lies furthest away from the optimum, and is therefore limiting, is determining for the survival and abundance of a species.

**1** River nerites

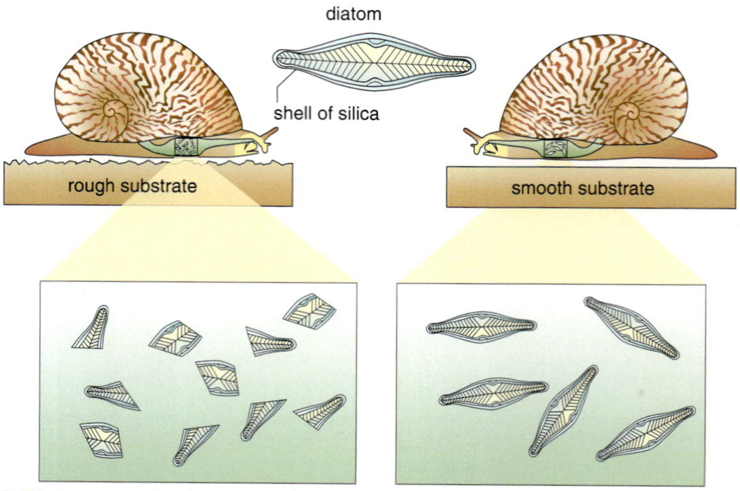


**2** Diatoms as a food resource for •iver nerites

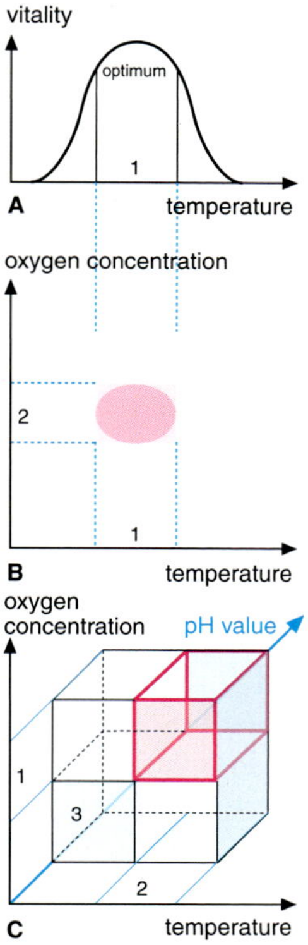


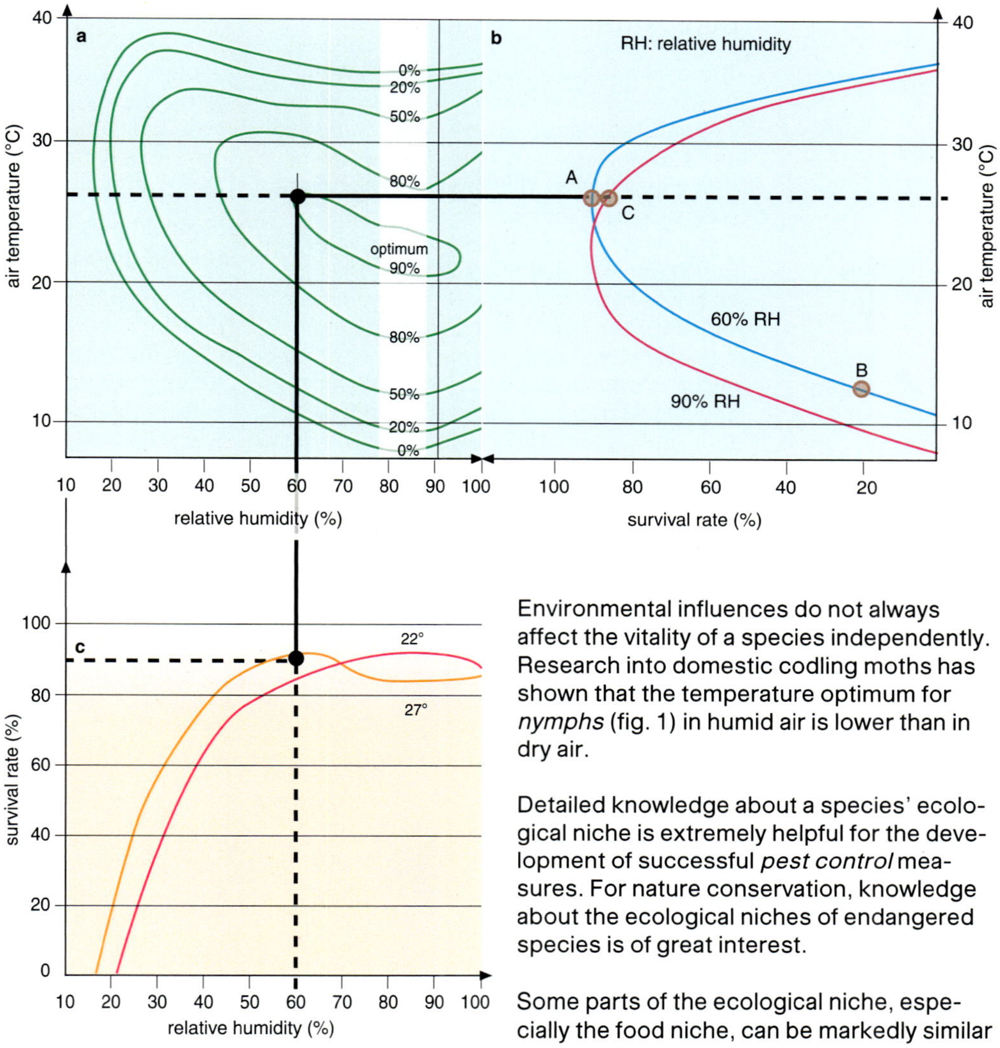


**1** Codling moth: two relevant factors

In agriculture this is known as "*Liebig's law of the minimum*". If only one important mineral is missing in fertilisers, the plants will not be able to grow optimally. The addition of this substance increases the yield of a harvest.

Some requirements can only be described in a qualitative way. Many other factors such as temperature, pH value and humidity can be measured quantitatively as ecological potencies. Up to three of such potencies can be displayed in a single graph (see margin). The common intersection, i. e. the overlap of all three factors, signifies the requirement of that species with regard to all three factors. The total of all ecological potencies describes the ecological niche quantitatively; it is however impossible to display this in a graph.

mole
(Soricoidea, Talpidae)

golden mole
(Soricoidea, Chrysochloridae)

marsupial mole
(Marsupialia, Notoryctidae)

Environmental influences do not always affect the vitality of a species independently. Research into domestic codling moths has shown that the temperature optimum for *nymphs* (fig. 1) in humid air is lower than in dry air.

Detailed knowledge about a species' ecological niche is extremely helpful for the development of successful *pest control* measures. For nature conservation, knowledge about the ecological niches of endangered species is of great interest.

Some parts of the ecological niche, especially the food niche, can be markedly similar in highly different parts of the world. In such cases we speak of *equivalent positions*. On all continents, for example, we find small mammals similar to the mole; they live mainly underground and also find their food there. Adaptation to similar environmental conditions results in surprising similarities between unrelated species as can be observed between the star-nosed mole (North America), the marsupial mole (Australia) and the golden mole (South Africa).

## Tasks

① Explain the data points A, B, and C in diagram 1b by using diagram 1a.

② Use the Internet to compare the different temperature optima at different humidities for the nymph of the codling moth with the living conditions of the moth.

## Competition for resources

The *common buzzard* makes a variety of demands on its environment. Theses birds need an open landscape with plenty of mice and elevated lookout places from which they can start their hunting flight. High trees are needed for their nests. Some climatic factors are also of significance: cold and humid weather in spring increases the mortality rate of their young, and heavy snow in winter makes hunting for mice almost impossible. Forces that have a *non-reciprocal* impact on the buzzard (e.g. temperature and snow cover) are called *ecological factors*. Interdependence does however exist between the food (mice) and the buzzard. By hunting mice, the buzzard reduces their numbers — the buzzard has an impact on this factor by depleting it. These parts of an ecological niche are called *resources*. Some factors occasionally fulfil a dual function: for aquatic organisms, water is the medium in which they live and is therefore an ecological factor. When water is taken in, on the contrary, it is a resource that is expended.

For *carnivores* such as buzzards, food is an especially important resource as it is limiting for their population. If food is available in abundance, other factors can become limiting, e.g. the amount of sheltered space for the rearing of offspring. *Seabirds* that breed on steep cliffs are a well known example. The sea provides an ample source of food, especially since they can fly long distances with comparatively little energy input. However, the number of *breeding areas* on the cliff wall is limited (see margin).

### Competition between the members of one species (intraspecific)

When resources are in short supply, *competition* occurs between all organisms using this resource. This is especially true for other members of the same species. This *intraspecific* competition often leads to the *marking of territories*. In order to ensure their share of a limited resource, an organism can claim and defend a defined area against members of the same species (*conspecifics*). As defending a territory demands considerable amounts of energy, a territory is usually only just big enough to ensure a sufficient supply of the resource in question. Where there is a high prey density, the territory of buzzards will be smaller than where prey is scarce. Often territories are only defended for a limited time during the year. As food is not a limiting resource for sea birds, the marking of territories on the sea would not make sense in a biological way — let alone the problem of clear border definition. These species however invest considerable energy in defending their small space on the cliff wall, such ensuring a scarce resource, their breeding area.

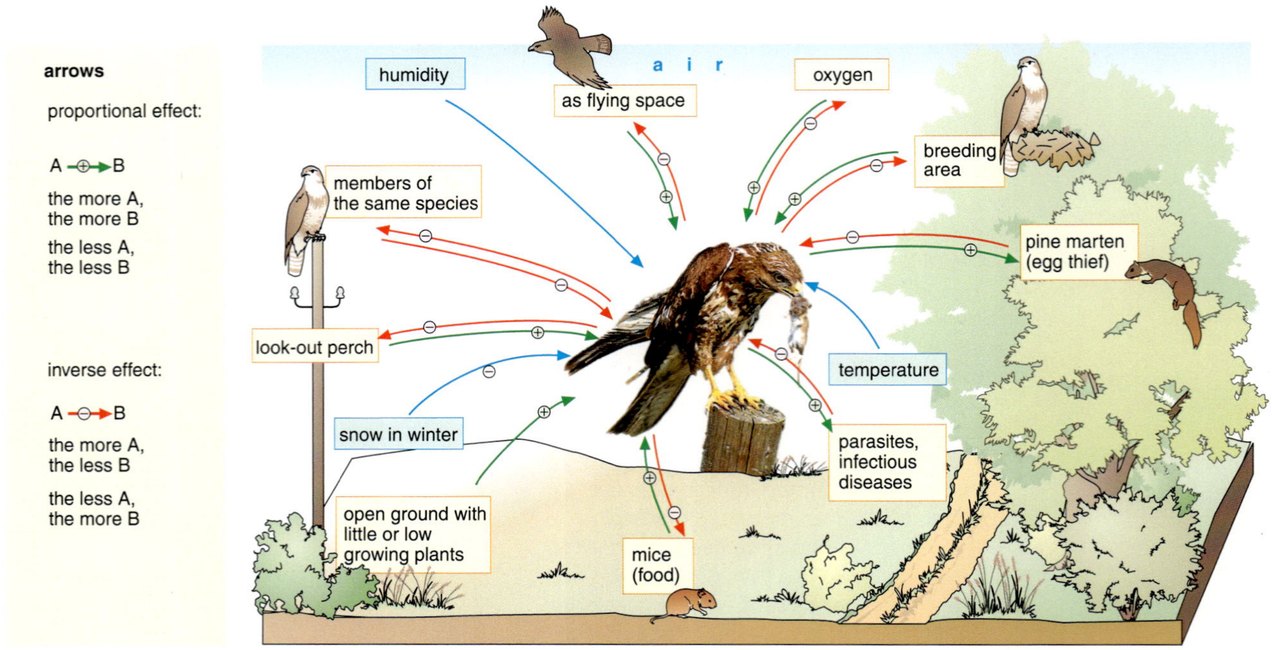


1 Ecological factors and resources of the common buzzard

**1** Himalayan balsam

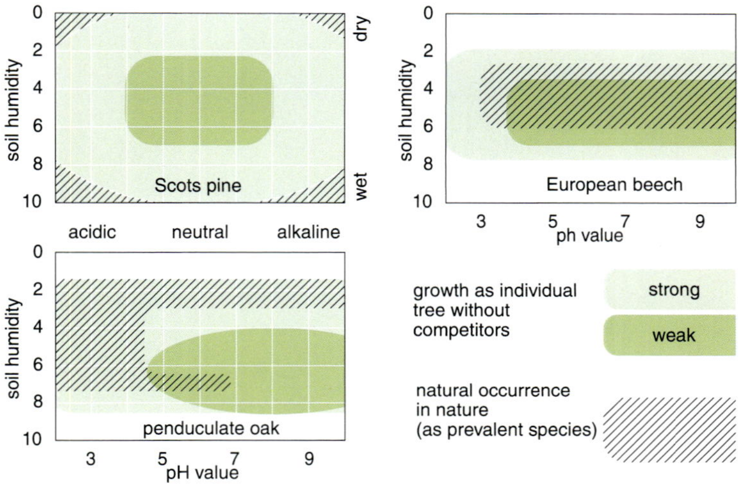


**2** Forest trees: fundamental and realised niches

## Competition between members of different species (interspecific competition)

*Himalayan balsam* used to be a popular garden flower in the 19th century; today it has turned into an invasive weed. This plant, originally from the Himalayas, is about 2 m in height and is mainly spreading near rivers and lakes, displacing local plants. This is possible as it is a strong competitor; it grows quickly and overgrows other plants. In addition to its making the best use of the resources of light and rooting space, the real strong point is mainly its production of high-quality nectar. This characterises the *competitive strength*. Bees and bumblebees quickly learn that the flowers of the himalayan balsam offer more nectar than the flowers of local plants. Being preferred for pollination ensures an even faster spreading of the plants. Near lakes and rivers, the himalayan balsam poses a particular danger. As the plants and their roots die in autumn, the decomposing roots cannot provide support for the soil and so an increase in erosion occurs during the winter months.

Highly competitive *neophyte*s such as the himalayan balsam are evidence for one important rule of ecology, the *competitive exclusion principle*. Different species in a habitat cannot coexist in the long run if their ecological niches are similar with regard to resources that are in short supply. The positive inversion of this rule (*avoidance of competition*) is exemplified by *seabirds*. These species show distinctive differences, especially with regard to limited resources, i.e. breeding area. The *black guillemot* breeds at the bottom of cliffs near the water's edge; *guillemots* position their eggs on protruding rock ledges and the *common shag* uses slightly larger plateaus on the cliff wall. Their differing use of this resource allows them to *coexist*.

## Fundamental niche and realised niche

*Forest trees* are another example of competitive strength. The *European beech* shows the best use of the resource of light and forms very dense tree tops. Competitors from other species cannot coexist here, whereas beech saplings can. In natural surroundings, the Scots pine, which has a high tolerance but is not a strong competitor, can only grow in places where factors such as humidity or pH value are not ideal and which are therefore not populated by other, less tolerant trees. The actual occurrence in nature defines the realised niche. For species that are not competitively strong, this realised niche is much narrower than the fundamental niche, which summarises the requirements of a species under experimental conditions without the influence of biotic factors such as competition or predation. The fundamental niche is determined by physiological factors, whereas the realised niche is defined by interactions with other species.

### Tasks

① Investigate, by using figure16. 1, to what extent factors with reciprocal impacts can be considered as resources.

② Compare the fundamental niche and realised niche for forest trees (fig. 2).

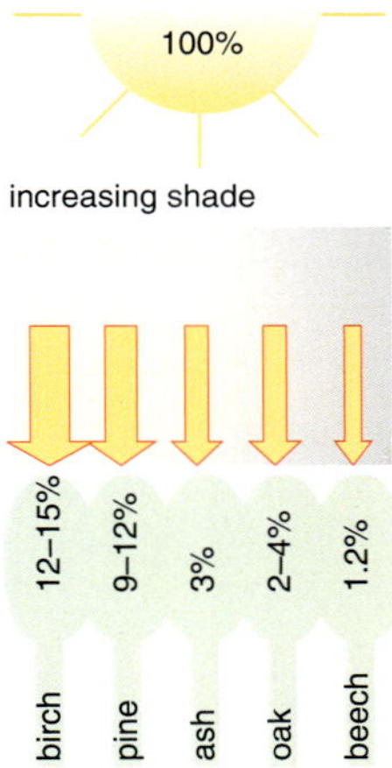


Minimum requirement regarding light of saplings/young trees

**Light incidence on the ground in a dense:**
forest of European beech approximately 1.6%
forest of oak approximately 4.0%

**Competitive exclusion principle**
Different species in a habitat cannot coexist in the long run if their ecological niches are too similar with regard to resources that are in short supply.

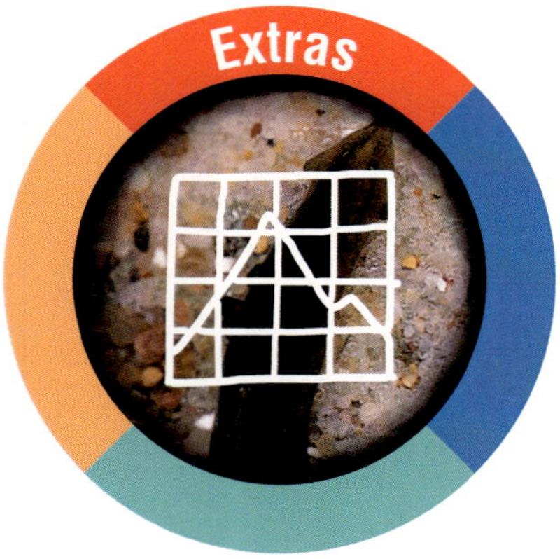


## The ecological niche of flatworms

Flatworms show a clear link between their occurrence and their environmental requirements. All species mentioned here are about 2 cm in length and mainly inhabit flowing rivers; they move by crawling along the rocky substrate in the water currents. They avoid light and have a good *olfactory sense*. They feed on dead animals and small invertebrates by putting their body over their prey, which ends up under the opening of the digestive tract at the bottom of the flatworm. The flatworm's success rate is determined by its ability to move. Flatworms do not have a circulatory system. Their flat body shape minimises the distance across which gases have to diffuse.

### Stream system A

Biologists found three types of flatworm in a river in a German low mountain range. Surveying and mapping (fig. 1) showed that the three species populated three different sections of the river.

Experiments in the laboratory provided data concerning the oxygen consumption of the animals at rest and their maximum crawling speed (fig. 3).

#### Tasks

① Determine the temperature ranges in which the three species occur in the German river.

② Use an arrow diagram (see p. 16) to show the links between oxygen consumption (fig. 2), crawling speed (fig. 3) and temperature.

③ What predictions can be made with regard to the temperature tolerance of all three species?

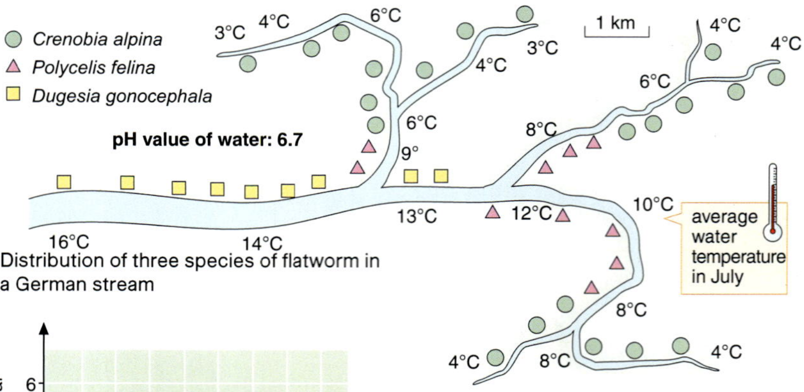


**1** Distribution of three species of flatworm in a German stream

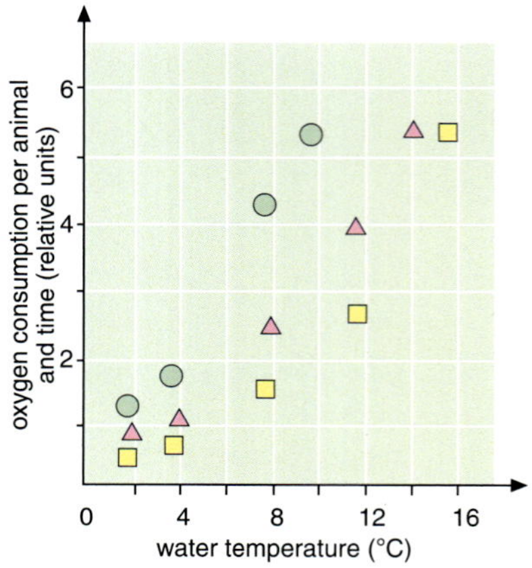


**2** Oxygen consumption

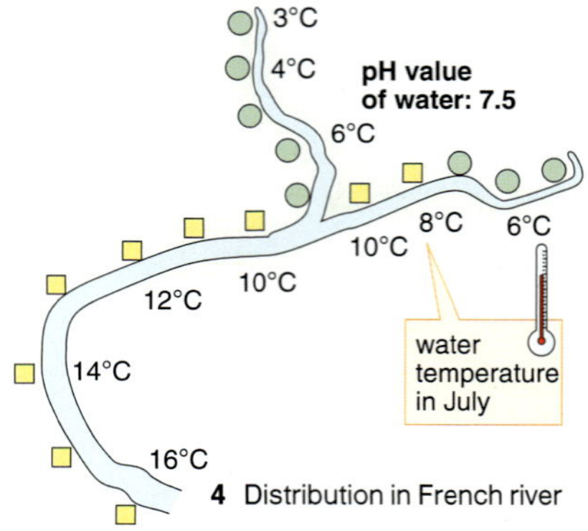


**4** Distribution in French river

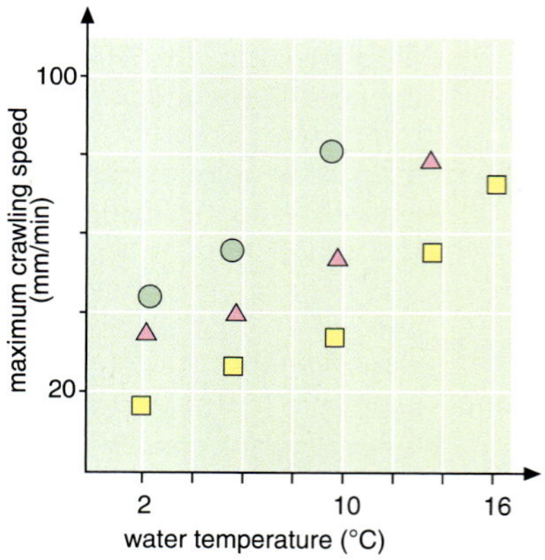


**3** Maximum crawling speed

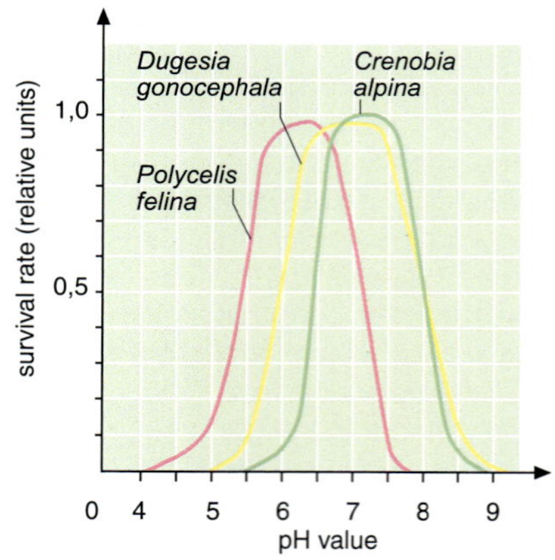


**5** pH tolerance of the three flatworm species

④ Propose a *well-founded* hypothesis and give reasons for the spread of these three species in the German stream.

### Stream system B

Research into a river of a similar structure in France produced a surprise: although *Polycelis felina* was missing over the entire stream system, it was nonetheless populated by flatworms throughout. The pH value in the French river was different from the stream in Germany.

#### Tasks

⑤ Give potential reasons why *Polycelis felina* was missing in France.

⑥ Compare the distribution of the species in both rivers and provide a possible explanation for the differences.

⑦ Suggest experiments that could be used to verify your hypothesis (task 6).

⑧ "For competitive species the fundamental niche is almost identical to the realised niche". Assess to what extent this statement is true in this case.

# Investigations into ecological niches

Research into light and temperature requirements of invertebrates can be performed both in the laboratory and outdoors. As only comparative studies are of relevance, at least two related species should be investigated, for which snails are a suitable subject.
It is sensible to compare the requirements of slugs versus snails.

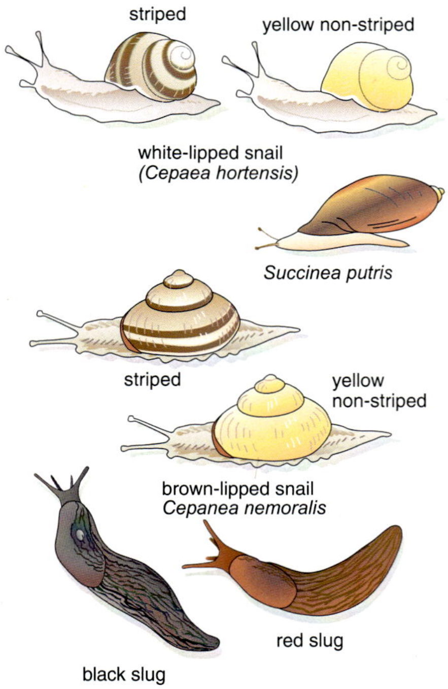


## Outdoor investigation

Choose two species that occur frequently in the area of investigation. (You will need at least 10, and ideally 20 individuals per species).

### Location

Choose two locations lying in close proximity and showing a difference in temperature:
- Hedge: south and north side plus weeds
- Edge of a forest: sunny/shade

Choose an area large enough to allow you to find sufficient individuals of each species.

### Equipment

- thermometer
- light meter
- at least 30 containers to store animals (numbered with blank label)

### Experiment

a) Draw a sketch of the area (to scale). Add elevated areas such as trees and shrubs.
b) Find individuals of the chosen species in both locations. In each location where an animal is found, measure the temperature and light intensity.
c) Put the animals into the containers and label them with the data measured. Add the number of the container to your sketch of the area.

### Tasks

① For each location and for each species draw up a grid/diagram:
- y-axis (vertical): light intensity (one data point for each animal)
  x-axis (horizontal): temperature
- Number of individuals of a species depending on light intensity

② Compare the two species with regard to their demands on their environment.

## Laboratory investigation

The temperature gradient should be set up on a "*temperature gradient plate*" (see diagram top right) three hours (or at least one hour) before the start of the experiment (NaCl/ice-mixture, boiling water) and the temperature should be measured at four different points. Ideally, there should be a linear gradient between 5 °C and 45 °C. The metal plate is covered with a glass plate.

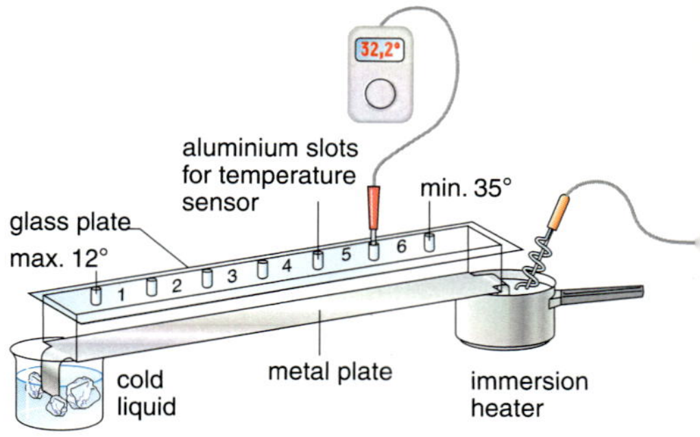


Animals of both species are placed in the area between 20 °C and 30 °C. Animals that show no activity can be put into lukewarm water for 30 s.

### Tasks

③ Count the distribution of animals on the gradient every 15 minutes. Once the numbers remain constant, draw a graph to show the results (*frequency distribution*).

④ Calculate the average temperatures for both species.

⑤ Compare your results with the data that you obtained in your outdoor investigation.

⑥ Does a species prefer light or shadow? Draw up a plan for a simple experiment and carry it out.

## Investigation with woodlice

Other invertebrates such as isopods can also be objects of comparative studies.

### Task

⑦ Make changes to the experimental set-up described above in order to be able to investigate the ecological requirements and preferences of different isopod species with regard to temperature und air humidity.

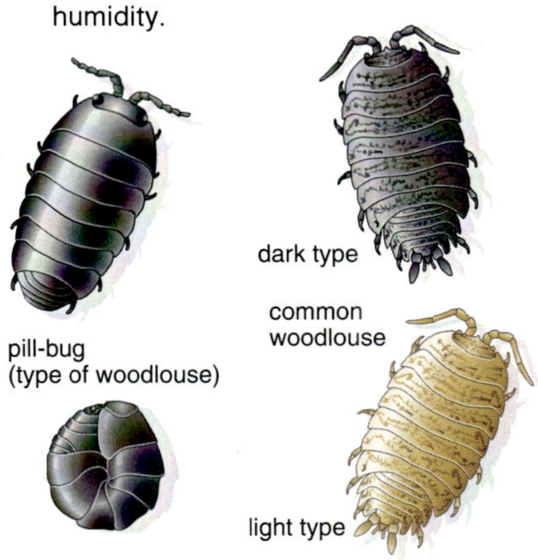

| Country | 1998 | | 1999 | | 2000 |
|---|---|---|---|---|---|
| **Brazil** | | | | | |
| Population size (millions) | 164.47 | | 167.27 | | 170.12 |
| Absolute growth per year (millions) | | 2.80 | | 2.85 | |
| Growth rate per year | | 0.017 (1.7%) | | 0.017 (1.7%) | |
| **Congo** | | | | | |
| Population size (millions) | 48.26 | | 49.80 | | 51.39 |
| Absolute growth per year (millions) | | 1.54 | | 1.59 | |
| Growth rate per year | | 0.032 (3.2%) | | 0.032 (3.2%) | |

**1** Population development of two countries

»info box«

### Model of exponential growth

The mathematical definition of the growth of populations is based on the following definitions:

$N_t$ = population size (number of individuals) after t time periods
r = rate of growth, percentage growth of population per period of time
The rate of growth is the difference between birth rate (b) and death rate (d): $r = b - d$

The new population size ($N_1$) is the sum of the previous figure ($N_0$) and the increase/change: $N_1 = N_0 + \text{increase}_0$

The increase is relative with regard to the current figure:
$\text{increase}_0 = N_0 \cdot r$
Putting both equations together: $N_1 = N_0 + N_0 \cdot r = N_0 \cdot (1 + r)$

For the two subsequent points in time this means that:
$N_2 = N_1 \cdot (1 + r) = N \cdot (1 + r) \cdot (1 + r) = N_0 \cdot (1 + r)^2$
$N_3 = N_2 \cdot (1 + r) = N_0 \cdot (1 + r^2 \cdot (1 + r) = N_0 \cdot (1 + r)^3$

In general, it can be said that $N_t = N_0 \cdot (1 + r)^t$

This is the general equation for exponential growth.

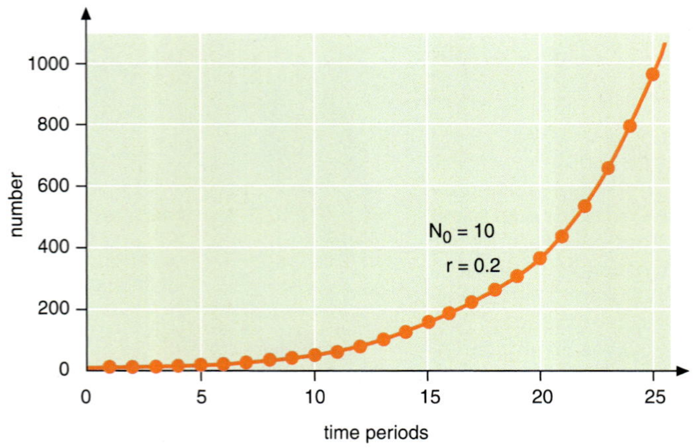


## Growing populations

In any given population, the number of individual members is continuously changing through births, deaths, and migration. Thus populations either grow or shrink. Infinite growth is obviously impossible because of the limitation of finite resources; this is also true for humans. It is of importance for all countries to judge their population development in order to take measures to be able to sustain its inhabitants in the future.

A comparison of two populations over several years will reveal different growth rates. Changes in absolute numbers are of minor significance as they always refer to different base values. Data concerning relative changes are more useful; these are called *growth rates* (fig. 1). In many countries, the growth rate of the human population is between 0.01 (1%) and 0.03 (3%); in some countries such as Germany, the growth rate is negative, i. e. the population is decreasing.

### Exponential growth

In order to calculate future population development, such data is used to create mathematical models that take account of the underlying principles of population development. The model of *exponential growth* provides a good first *approximation*. To keep things simple, the model does not take immigration and emigration into account. It furthermore assumes unlimited growth and uses constant birth and mortality rates, i. e. an overall constant growth rate. Exponential growth is the simplest way of modelling population development.

Flour beetles held under laboratory conditions show that population growth comes to a halt when resources become scarce; the number of beetles, i.e. the *population density* (number of individuals per area) reaches a plateau (fig. 21.1). This does not correspond to the model of exponential growth that, as one of its simplifying features, assumes unlimited growth. The model of exponential growth consequently needs to be replaced by a more sophisticated model.

**1** Growth of a flour beetle population under laboratory conditions

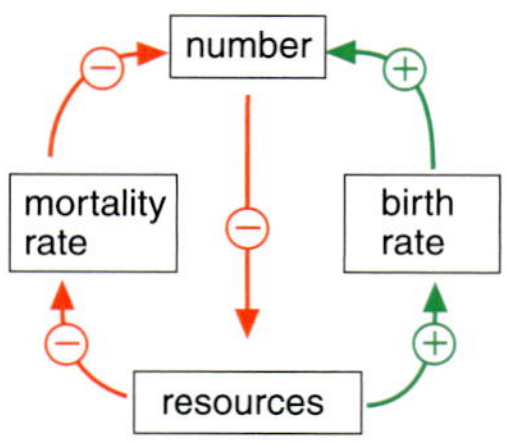


## Logistic growth

The growth rate is considered to be constant in the exponential model. However, as the experiment with flour beetles shows, any representation of growth rate needs to be variable and needs to decrease with increasing population. Observations of real populations show that the growth rate largely depends on the supply of resources. It is generally correct to assume that, with an increasing population size, the share of any resource for the individual becomes smaller; this results in lower fertility. Thus birth rates decrease, while the death rate goes up. Populations grow as long as their birth rate is higher than their death rate. Once the number of individuals reaches a certain limit, however, birth rate and death rate are equal and the population stops growing. This limit is called the *carrying capacity*; it denotes the maximum number of individuals that can live in a habitat. Once birth rates are lower than death rates, the population is on the decrease. The model that makes these assumptions is represented by the *model of logistic growth*.

### Task

① Calculate the population size for both countries (fig. 20. 1) in the years 2015 and 2020. In order to simplify things, start using the data of the year 2000 and assume exponential growth.

## »info box«

### Model of logistic growth

The assumption of constant growth rates in the model of exponential growth turns out to be not realistic, as both birth and mortality rates and therefore the growth rate depend on the size of the actual number in the population (N). Accordingly, they need to be integrated in any model.

The larger the population size (N), the lower is the birth rate (b) and the higher is the death rate (d). In simplified terms, this can be described by two linear equations:
$b = b_0 + m_b \cdot N$ and $d = d_0 + m_b \cdot N$

(1) Now we are able to calculate the intersection (at $b = d$). The number at the intersection of both lines signifies the carrying capacity (K) of a habitat:
$b_0 + m_b \cdot N = d_0 + m_b \cdot N$

This can be changed into (with $r_0 + b_0 - d_0$):
$K = r_0 : (m_d - m_b)$ or $m_d - m_b = r_0 : K$

(2) The current rate of growth depends on the number ($N_t$):
$r_1 = b - d = (b_0 + m_b \cdot N_t) - (d_0 + m_d \cdot N_t)$
This can be changed into: $r_1 = b_0 - d_0 - (m_d - m_b) \cdot N_t$
According to (1), $(b_0 - d_0)$ can be replaced by $r_0$, and similarly $(m_d - m_b)$ can be substituted by $r_0$:K. It can then be deduced that:
$r_1 = r_0 \cdot (1 - N_t : K)$

(3) The equation for logistic growth is obtained by the substitution of r in the original equation of exponential growth by $r_t$.
$N_{t+1} = N_t + N_t \cdot r_0 \cdot (1 - N_t : K)$

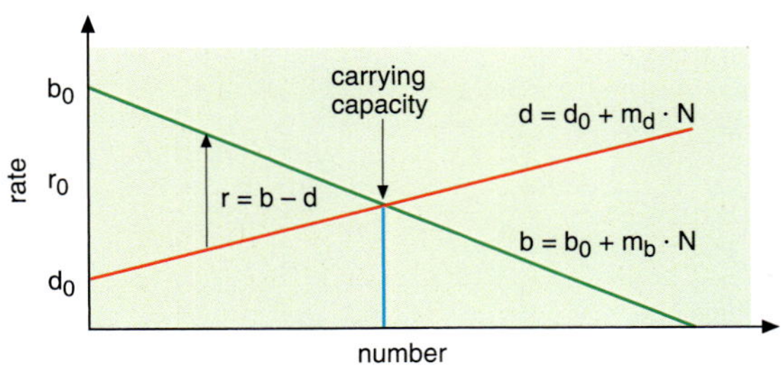


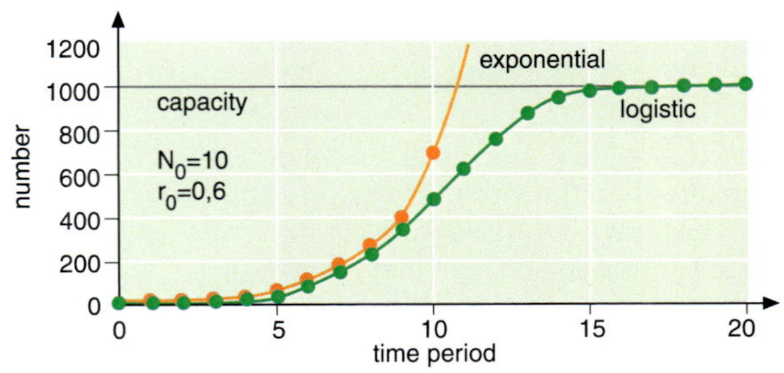

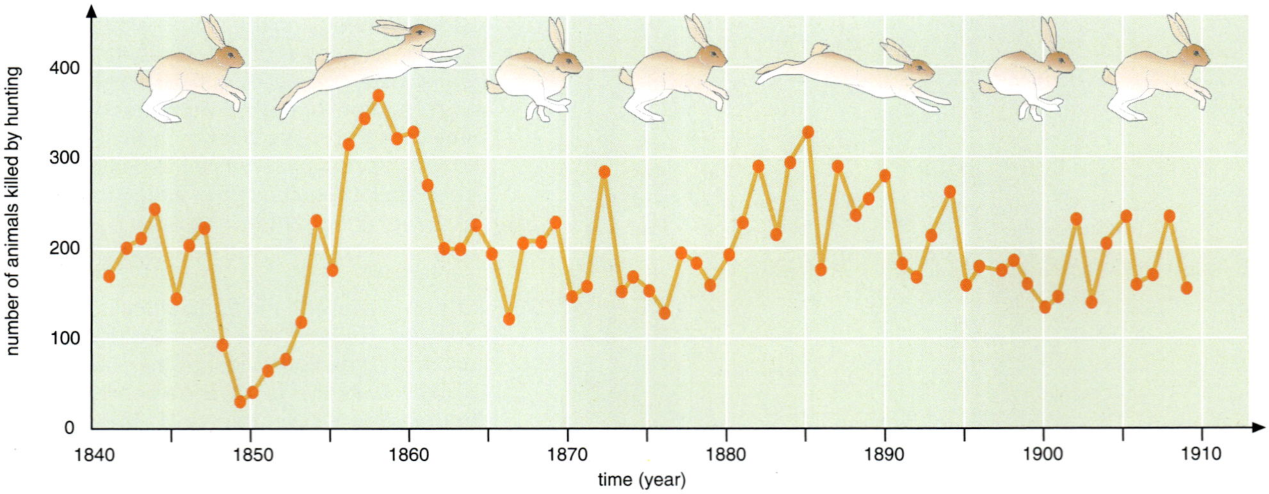


**1** Population of hares in the Principality of Liechtenstein (estimates based on annual hunting figures)

## Population sizes change

There are significant changes in the number of individuals of a population when observed over several years; hares are a good example (see fig. 1). Population sizes generally do not remain constant. The reasons for such changes can be grouped into *density-independent* and *density-dependent factors*.

### Density-independent factors

On the one hand, many individuals of a species do not survive because of catastrophes or adverse conditions that might occur. Many young hares, for example, freeze to death if the weather in spring is cool and humid. Bad weather also has an impact on the growing of the food resources of hares; this results in a more limited supply for all individuals of the population. In such circumstances, the population size does not increase and sometimes might even decline. A warm and humid summer on the other hand means excellent living conditions with an abundance of food resources; this then results in a large share of both the adult animals and their offspring surviving and consequently in significant population growth.

In these instances, the actual animal density has no effect on the size of the population – the reasons for the changes are external. Changes in population size that are density-independent are called population *fluctuations*.

### Density-dependent factors

Other, more regular changes in population size can be explained by the size of the population itself or by the size of other neighbouring populations (fig. 2). These more regular changes are called *oscillations* and can have a variety of reasons.

Individuals of a population might possibly use up their resources. During the time that it takes to regenerate the missing resource, the population collapses. Once the resource has grown back, the then diminished population can also recover and start the cycle anew.

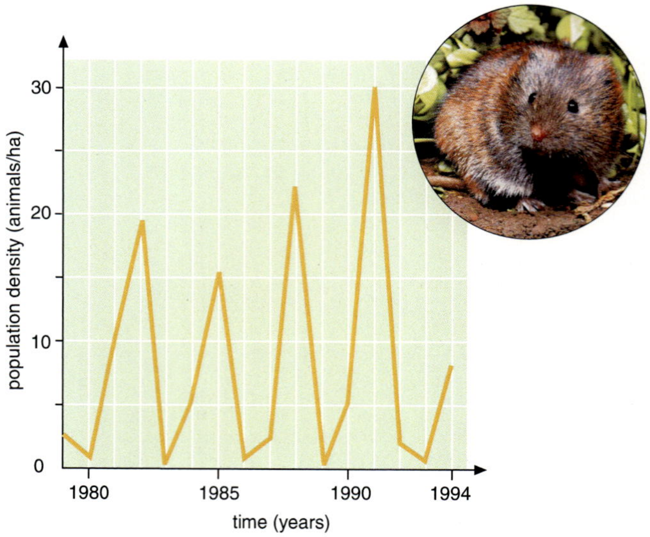


**2** Oscillation of population of European burrowing mice

The duration of one cycle depends on how quickly a resource can be regenerated and also on the ability of the population to survive times of scarce resources. Food and fresh water are resources for many organisms. If water becomes polluted through *faeces* or urine, it can become poisonous. Furthermore, the oxygen in these waters can be used up through processes of decay.

Infectious diseases can also trigger changes in population numbers. With increasing population size, the chance of catching a disease is higher. The rate of both infection and mortality go up and the population faces a collapse. Furthermore, high population density and the associated lack of resources can also cause stress, which can lead to aggression and infertility.

The analysis of changes in the numbers of natural populations shows that both density-dependent and density-independent factors are always acting at the same time and with different effects.

### Time-delayed effects of density

Individuals of some species can store resources and thus can reproduce, even when the actual conditions in the environment no longer provide enough resources. These populations seem to respond to the conditions at that point in time when the resources were placed in store, i.e. their response is time-delayed. They outperform their capacity and collapse thereafter. This can also happen in cycles and the population oscillations of the European burrowing mouse can be explained in this way.

**»info box«**

### Experiments with the water flea Daphnia investigating the reasons for population oscillations

Populations of *Daphnia magna* were kept in 50 $cm^3$ of water and regularly given the same amount food. The daily births and deaths were recorded and the population size was updated accordingly. Two populations were kept at different temperatures (18 °C and 25 °C). The results show clear differences between the two experiments.

After a phase of growth, the population at the lower temperature reached an almost constant population size of about 120 individuals. This growth corresponds to the model of logistic growth. Significant fluctuations occurred at the higher temperature, despite conditions being otherwise the same. Possible explanations can be found in the death and birth rate of these populations, which were also recorded.

The rate of metabolism is lower in colder environments (reaction rate-temperature rule or van't Hoff's rule). Consumption and the processing of resources take longer and so birth and death rates stay at constant levels and run almost in parallel; the population numbers remain almost constant.

In warmer environments, all metabolic processes speed up. The rapid intake of resources results in a brief increase in births and a lower mortality rate; these no longer run in parallel. The following phase of population growth leads to a higher rate of consumption, which finally ends in a lack of resources. The number of births goes down as a consequence and the mortality rate goes up; the population overall decreases in size. This seems to be followed by a third phase, during which bacteria possibly decompose the dead *Daphnia* in a cleaning up process, before the cycle of growth can start anew.

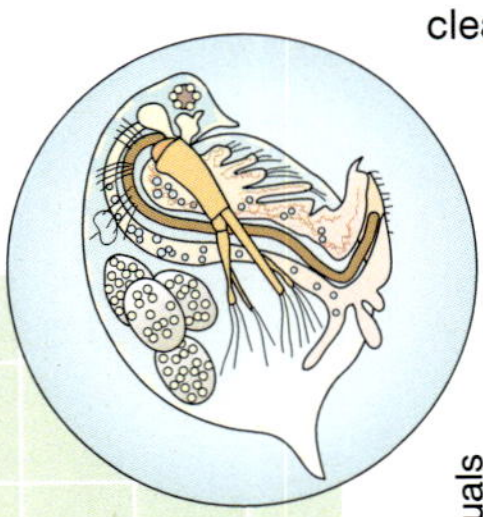

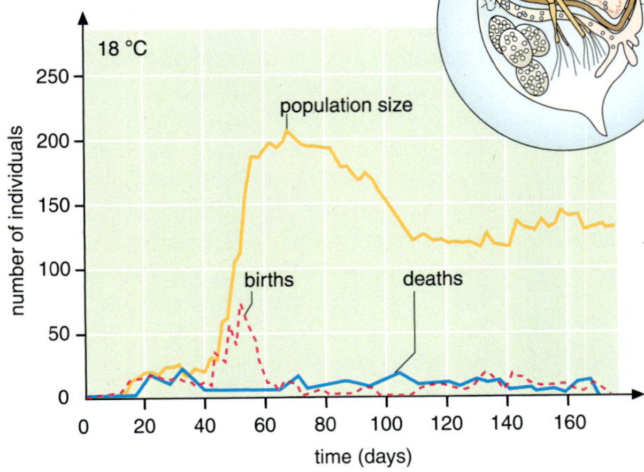


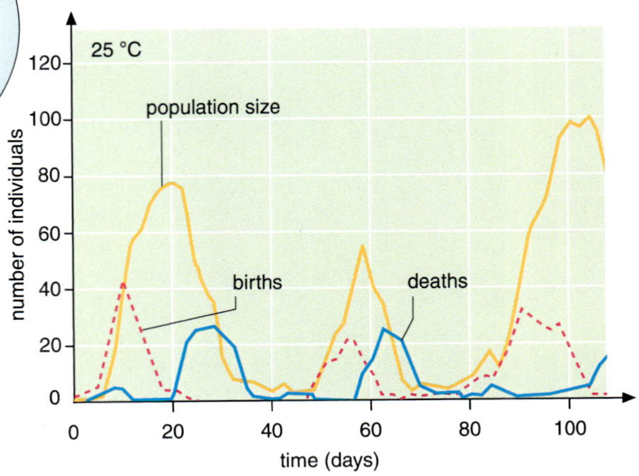

**1** Buzzard catching a mouse

**2** Predation strategies and defence mechanisms

## Predator and prey

Buzzards can be observed circling the sky over rural areas. All of a sudden they swoop down to catch a mouse with their claws. Mice are prey for the buzzards, whereas buzzards are the predators of mice. Buzzard and mice are an example of a *predator-prey relationship*.

For a predator its prey is a necessary food resource. Predators attack, kill or at least damage their prey and eat it completely or at least in part. For the prey, predators are a threat, affecting or endangering their survival.

The populations of predators and prey therefore influence each other. Many of the regular patterns seen in predator-prey relationships apply not only to examples such as the buzzard and mouse, but are also found between herbivores and the plants that they eat (*plant-herbivore relationship*). They even apply to the relationships between parasites and their hosts (*parasite-host relationship*); these also count as predator-prey relationships.

### Mutual adaptations

Predators have various hunting strategies; prey in turn develops mechanisms to avoid being hunted. Among predators there are, for example, hunters that actively pursue their prey. Among these are weasels, birds of prey, sharks, dragonflies and many other species. Other predators erect traps, as do spiders. Many aquatic predators are *filter feeders*. Among these are small animals of the zooplankton, such as *Daphnia*, together with mussels and also massive animals such as baleen whales. *Grazers* on the other hand search for their food and graze on it, e.g. cows or the caterpillars of many insects. Other herbivores such as squirrels collect seeds and fruit.

Prey such as antelopes can manage to escape by their speed or can remain unnoticed by camouflage or have a hidden way of life as many insects. Plants often contain bitter or toxic substances that protect them from being eaten. Both animals and plants sometimes have hard shells or thorns that make it harder to attack them.

**1** Locust swarm in Africa

brown tree snake,
*Boiga irregularis*

ineffective predator

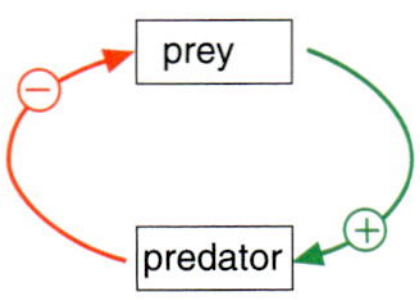


moderately effective

highly effective

## Interaction of populations

The interdependence of predator and prey changes the populations involved. When prey is readily available as a resource, individuals of the predator population have a better chance of survival, can grow stronger and can produce and bring up more offspring. If prey is scarce, there is a higher mortality and fewer or no offspring in the predator population. The number of prey determines whether the predator population can grow or declines. As predators remove individuals from the prey population, the latter decreases more quickly when more predators are around. The extent to which predators have an impact on the prey population depends on the predators' efficiency. Refuges or defence mechanisms diminish the predators' efficiency. High efficiency gives predators an advantage, whereas improved defence systems or places that are difficult to access help the prey. The conditions in the particular living space strongly impact on a species' efficiency. The predators' efficiency strongly influences the interaction of predator and prey.

**Less effective predators** hardly impact on the prey population. Mice, for example, mostly live hidden underground. They only spend short periods of time at the surface and can quickly disappear into their burrows. Of all the buzzards' hunting attempts, only 5 % are successful, which makes them ineffective predators, and they only eat a few of the mice living in their territory. In addition, buzzards only reproduce once in spring, which means that their population grows slowly, whereas mice can rear many generations of offspring every year. There is consequently a one-way impact by the prey on the predator.

**Moderately effective predators** lead to a reciprocal impact of predator and prey population on each other, much like a feedback system.
- More prey means a better food supply for the predators and therefore leads to increased reproduction rates.
- The predators now use up more prey and so reduce the prey population.
- Reduced prey leads to starvation of predators and their numbers decline.
- A smaller number of predators now enables the prey population to grow again.

This feedback system can result in fluctuations in both populations; sometimes, these fluctuations can subside and end up in a situation of equilibrium, where both populations retain an almost constant size.

**Highly effective predators** can strongly reduce the number of prey. This true for some herbivores, e. g. whose "prey" are plants. Grazing herds or migrating swarms of insects can completely raze entire regions (fig. 1). Plankton organisms such as *Daphnia* can regularly eat up all the algae in a lake (clear-water stadium, see p. 51). For the prey population, this is always a dramatic collapse and the predators cause the subsequent lack of resources themselves with which they cope by migration to different areas, a change of prey or the formation of special survival stadia (e. g. the winter eggs of *Daphnia*).

When predators are transported to regions in which they are not normally indigenous, they often prove highly effective and can entirely extinguish domestic species. In 1945 for example, the brown tree snake was brought to Guam, an island in the Pacific Ocean, and has strongly proliferated there. It feeds on small mammals and birds and has now exterminated some domestic species. Similar things have happened on islands to which sailors have brought rats or domestic animals such as pigs or cats.

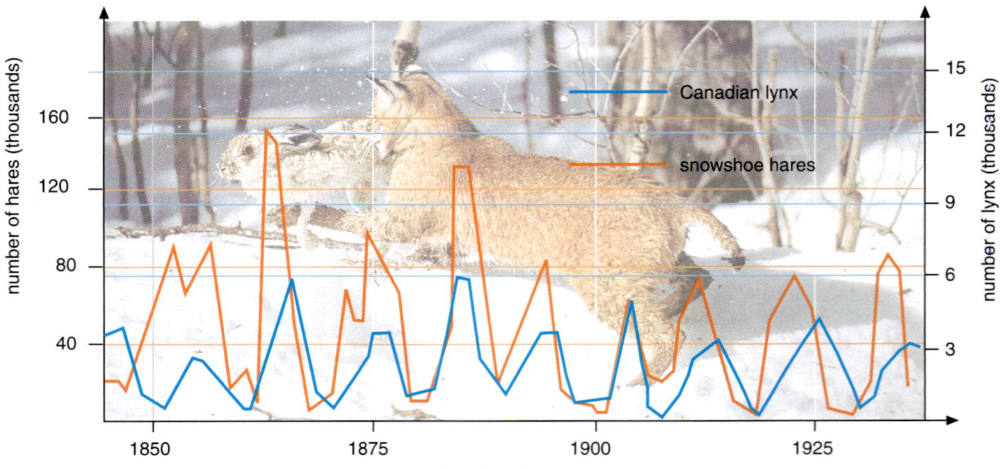

**1** Populations of lynx and snowshoe hare in Canada, counted in coats

## Modelling predator-prey relationships

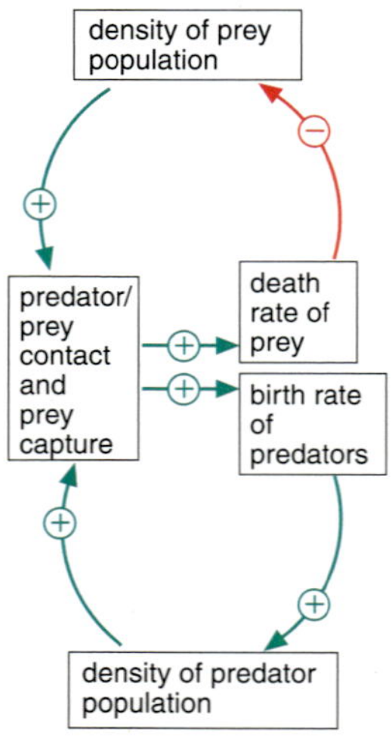


Many populations of predators and their prey grow and decrease and it is often easy to assume a simple kind of interaction. The counts of coats of lynx and snowshoe hares by the Canadian Hudson Bay Company show such changes and possibly exemplify such a case (fig. 1). Reciprocal impact can however rarely be established as the cause of population changes and so can only be applied to this single case. More detailed testing has shown that various populations are usually involved and that focusing on only two populations is over-simplistic.

Nonetheless, many attempts have been made to describe and understand the underlying principles, even if for this purpose the natural conditions need to be simplified first. One important way of doing this is to devise a mathematical model based on the hypothesised relationships. Scientists then try to use the outcomes of this model to explain their observations in nature.

Many models have been developed to describe predator-prey relationships. All models use separate equations to describe the growth of each population; these equations also take the influence of the populations on each other into account.

The most famous model was developed in the 1920s by the mathematicians ALFRED LOTKA and VITO VOLTERRA.

They tried to find a general explanation for the changes that they had observed in fish populations in the Adriatic Sea or with parasites. Their model is often applied to try and understand the system between lynxes and snowshoe hares.

LOTKA and VOLTERRA assumed exponential growth and varied the growth rate (r), which they obtained as the difference between birth rate (b) and death rate (d) (see p. 20).

$$N_{t+1} = N_t + r \cdot N_t = N_t + (b - d) \cdot N_t$$

In their model, the birth rate of the predator population only depends on the amount of prey. The more prey there is, the higher the predator birth rate. The death rate is considered to be constant. Correspondingly, the death rate of the prey population is assumed (only) to depend on the predator population. The more predators there are, the higher is the death rate of the prey population. The birth rate of prey is considered as being constant.

An investigation into the outcomes of the mathematical model indeed shows permanent fluctuations of both populations around their individual mean value, as can be observed with lynx and snowshoe hares; the model by LOTKA and VOLTERRA seems appropriate for the explanation of the changes in populations. From this model, some important rules can be derived, which are named after their discoverers (*Lotka-Volterra equations*):

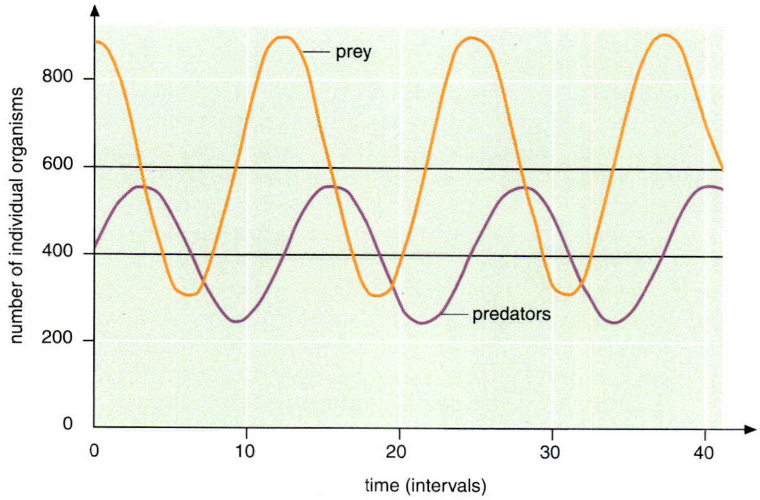


**1** Population changes according to the Lotka-Volterra model

- The mean values of both populations fluctuate periodically. The maxima and minima of the predator population follow the corresponding values of the prey population (First equation of LOTKA-VOLTERRA)
- Both populations fluctuate around a mean value that remains constant over long periods of time provided that environmental conditions do not change (Second equation of LOTKA-VOLTERRA)
- After a strong decrease in both populations, the prey recovers first. The recovery of the predator population follows thereafter (Third equation of LOTKA-VOLTERRA)

The model (fig. 1) shows the cyclic properties that can be observed with some natural predator-prey systems. However, this is still insufficient. It is, for example, based on the concept of exponential growth, i.e. without predators, the prey population would grow infinitely as there would be no upper limit. Furthermore, the death rate of the prey is only determined by the number of predators; without predators, the prey would not die. In addition, the model assumes the occurrence of cyclical fluctuations; virtually no time in the model is allowed outside one of these cycles. This can convey the wrong impression that the fluctuations as observed between lynx and snowshoe hares are necessary or characteristic features of any predator-prey system. In reality, however, only rare cases exhibit such fluctuations.

Many other models have been proposed based on *logistic growth*. One basic idea is that the potential number of predators depends on the amount of prey available. Based on this assumption, the capacity of the predator population is dependent on the amount of prey and can be integrated into this model. With regard to the prey population, the number of prey consumed by the predators can be taken as an additional loss and can so be integrated into the logistic equation of the prey population. Models such as these also display the properties defined by the equations of LOTKA and VOLTERRA but they no longer have the *flaws* mentioned previously.

»info box«

## Descriptive models and conceptual models

### Descriptive models

Many phenomena in nature are hard to imagine, especially when they lie outside our everyday experience; this is why we use models to gain better insights into such phenomena. Models of cells, for example, visualise the structure of and relationships between cell organelles. Functional models reduce structures to their absolute essentials to show how things work. Models of the wing movements of insects or models of the function of the diaphragm during breathing are examples. In theses cases, models are derived from nature and try to describe known facts; this is why these models are called *descriptive models*.

### Conceptual models

On the other hand, situations often occur in biology when scientists try to find a unifying basic principle or law that is common to a variety of individual and seemingly unrelated cases. Then it is often useful to devise scenarios and to summarise them in a mathematical model. These are of course derived from ideas and assumptions gained from biological observations and are not intended to be factual, but are proposed in order to allow access to unresolved questions. The analysis of the outcome of such models can provide hints to possible relationships in nature. These models therefore lead the way back to nature and are called *conceptual models*. Like the model of LOTKA and VOLTERRA, most models regarding populations are conceptual ones.

# K- and r-selection

**Strategy**
Structure or patterns of behaviour that allow the survival of a population

Nature-loving gardeners who want to have a colourful *meadow of flowers* on their plot of land might choose a wild flower seed-mix which is commonly available in gardening and do-it-yourself shops. After a few months, they will indeed be able to enjoy a colourful meadow of wild flowers. However, disappointment strikes in the years to come: the show of flowers goes down, the meadow is more green than colourful during most of the year and flowers no longer predominate.

A flower seed-mix contains mainly *annual flowers*. They are characterised by investing almost all of their biomass in their reproductive organs, i.e. their flowers, and later in their seeds. In the long run, they become replaced by *perennial herbs* such as grasses or cow parsley. These plants are stronger competitors as they invest energy in their *storage organs* rather than in their flowers; this enables them to survive during the winter months. In spring, their shoots are stronger than those of annual plants that only have small amounts of stored *nutrients* in their seeds.

Organisms that invest most of their biomass into reproduction are called r-strategists. They produce many *offspring,* but invest less of their energy into strength and permanence. They are characterised by being short-lived. Their mortality is high during every stage of life; their populations show high growth rates and fluctuate frequently. In contrast, organisms that direct a large amount of their biomass into securing their own existence and survival (through food storage, resistance against predators and environmental impacts) become stronger competitors and live longer because of this *investment*. The population size of these species is usually close to the limits of their capacity (K), which is determined by resource availability. This is why they are called K-strategists. The amount of energy available for a living being is limited. Investing energy into structures that increase competitive strength results in investing less energy into reproductive organs and therefore the number of offspring.

**1** Meadow of mostly annual flowers

**2** Perennial meadow

The ideal strategy for the survival of a population depends strongly on the *constancy* of its living conditions. A lake with constant living conditions will contain mostly K-strategists such as fish. A *puddle* on the contrary only exists for a short period of time. It is populated by r-strategists such as *unicellular* organisms and water fleas. Human impact can also be a determining factor for the ideal strategy. The high production of seeds by annual plants allows them to repopulate clear-cut areas quickly after deforestation; this living space has become free from competition because of human disturbance. If this area remains undisturbed for a long time, trees will succeed as they are extremely strong K-strategists and will eventually produce a forest once again (see p. 42).

**1** Koala

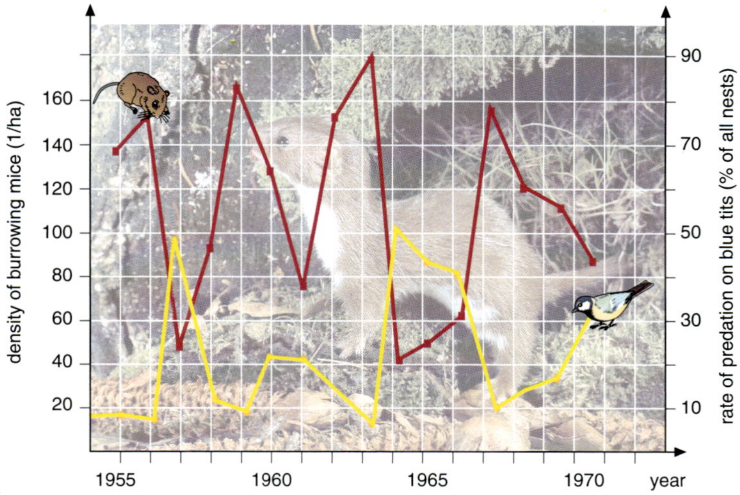


**2** Weasel – changes of prey

## Food strategies: specialists and generalists

In Australia, large *eucalyptus forests* can be found in areas that are not too dry. Because of its toxic cyanides and other compounds, these plants are well protected from predators. Nevertheless, one marsupial exclusively feeds on eucalyptus leaves: the *koala bear*. Through a number of adaptations, the koala is the only mammal that can use this plant as a food resource. The eucalyptus leaves are digested by special bacteria in the 2-m-long *appendix* of the koala (see margin). These microorganisms also manage to break down the poisons into harmless substances. Young koalas obtain these bacteria by eating special *faeces* of their mother. Koalas spend one fifth of their daytime eating, almost the rest of the time they are sleeping. Their movements are extremely slow. All these energy-saving behaviour strategies are adaptations to the small energy content of eucalyptus leaves.

The koala bear is a *specialist*, i.e. they are experts in using a certain resource; on the other hand they completely depend on this resource as they cannot replace eucalyptus by anything else. This specialisation is only possible as their food is available all year round and in great abundance. In our domestic fauna we can find similar examples. Herbivorous caterpillars of butterflies are often specialists. Plants that are toxic for other animals can be used by specialists through physiological adaptations.

The *weasel* on the contrary is a *generalist* from central Europe. Its prey-spectrum comprises mainly burrowing mice, but also other types of mice, birds and bird eggs (mainly those of blue tits). Generalists are *opportunistic*, i. e. they choose their prey so that they only need to invest a *minimum input of effort* and, at the same time, can make the greatest use of their energy. If the burrowing mice population decreases, the weasel will prefer other types of prey (fig. 2). This *change of prey* has consequences: the populations of predator and prey are less interdependent than they are with specialised predators. Constant population sizes can be expected if no other factors, such as the climate, become limiting. In north Finland, the only food resource available to the weasel are burrowing mice. However, the weasel is extremely well adapted to this type of prey: with its slender body, it can follow and hunt the burrowing mice in their underground tunnels or under snow. In Finland, the weasel is a *food specialist* and therefore highly dependent on prey density.

### Tasks

① Interpret the data regarding the choice of prey by the weasel (fig. 2).

② Relate the concepts of stenopotent and eurypotent to generalists and specialists.

③ Specialists virtually never exterminate their prey. Explain this statement.

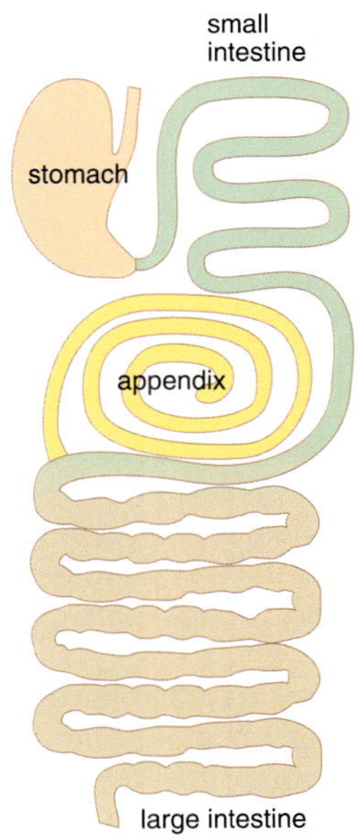


digestive system of the koala

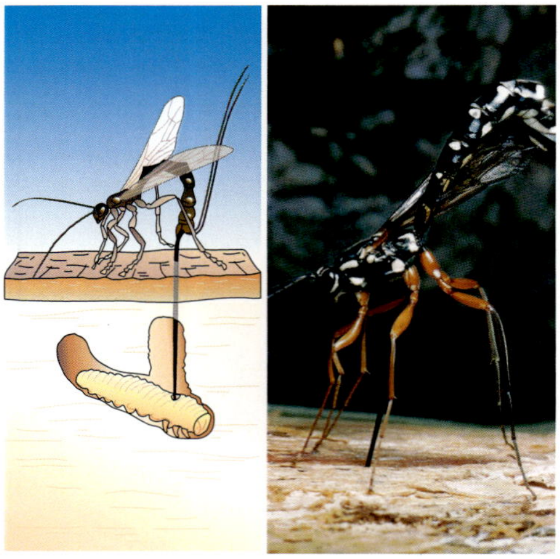

**1** Ichneumon wasps

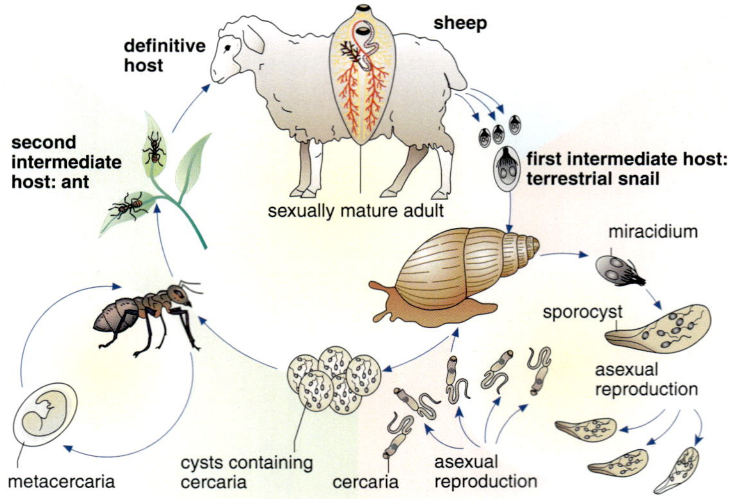


**2** Stages of development of the lancet liver fluke

## Parasitism

Predators kill their prey before eating it. *Parasites*, on the contrary, feed on the other living beings (*hosts*) without killing them. This damages the host, resulting in a shorter life-span, less growth or a lower reproduction rate, but does not necessarily lead to the host's death.

Unlike most predators, parasites are much smaller than their hosts, which is why the growth rate (r) of the parasite population is significantly higher than that of the host population. Parasites therefore often have a stronger impact on the host population than predators have an impact on a prey population.

Parasitism occurs in numerous ecological variations. Parasites can live on the surface of a host (*ectoparasites* such as *fleas, tick* or *head lice*). This type of parasite typically has special organs to attach itself to the host; it also has a flat body shape (see central column). Some animals can be considered as having characteristics in-between parasites and predators (*parasitoids*); the larvae of ichneumon wasps are specialists and populate caterpillars of butterflies (fig. 1). At the end of their larval stadium, the host dies and the *imago* leaves its host.

Being small allows parasites to live inside their hosts (*endoparasites*). Because of defence strategies of affected host organisms, parasites must be extremely well adapted to their host; this is why endoparasites are usually highly specialised.

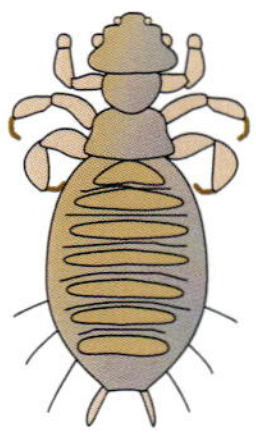

head louse

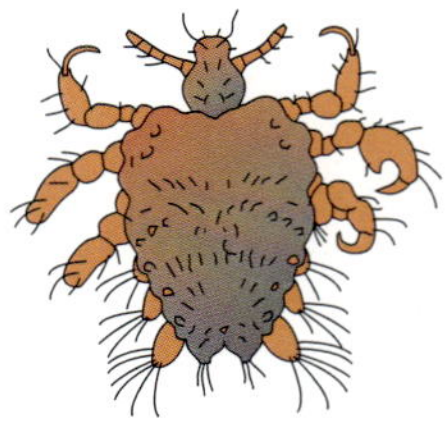

crab louse

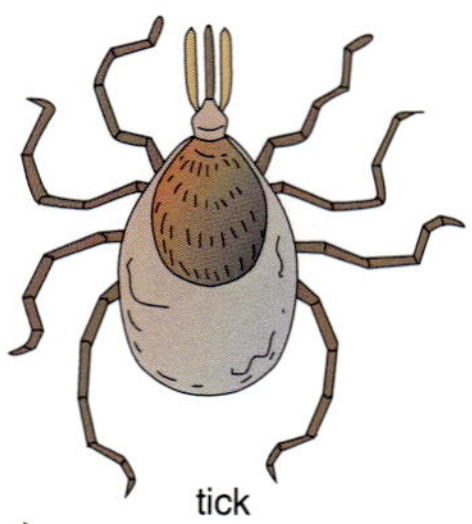

tick

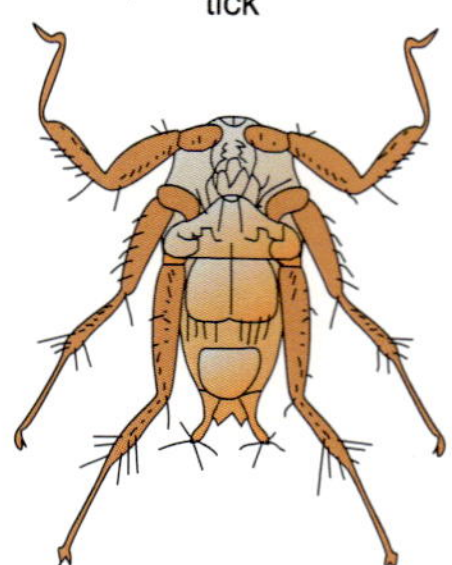

bat fly

The probability of a host becoming directly infested by a parasite is low. The example of the lancet liver fluke (fig. 2) demonstrates its adaptations that ensure its survival: this parasite is a *hermaphrodite* that lives in the liver of sheep. Like many other animals, sheep avoid contact with their faeces, which is why a direct transfer from one animal to the other is unlikely. Certain snails however feed on the faeces of mammals; this makes it highly likely that the parasite eggs will end up in this *intermediate host*. *Cysts*, which are excreted by the infected snails, contain the parasite in its next developmental stadium together with many carbohydrates. As they contain a large amount of energy, they are a popular food source for ants. The behaviour of the ants that become infected in this way is extraordinary: they climb to the top of a blade of grass and clamp their *mandibles* into the grass and stay there for days, increasing the probability of being eaten by sheep which then again become the *definitive host* of the parasite; the wheel has come full circle. The *change of hosts* is therefore a strategy of the parasite to enable it to reach new hosts. Various *asexual reproductive cycles* increase the probability that at least some parasites end up in a new definitive host.

### Task

① Find information about pathogens such as bacteria causing, for example, bubonic plague or salmonella. Do they show typical features of parasites?

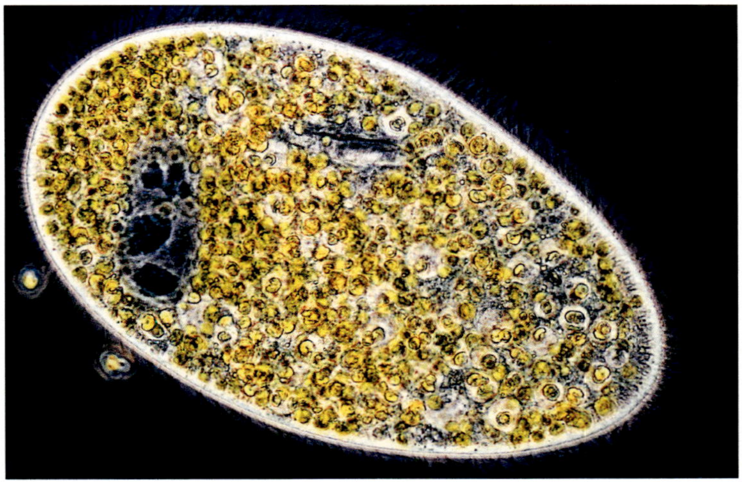

1 *Paramecium bursaria* containing *Chlorella*

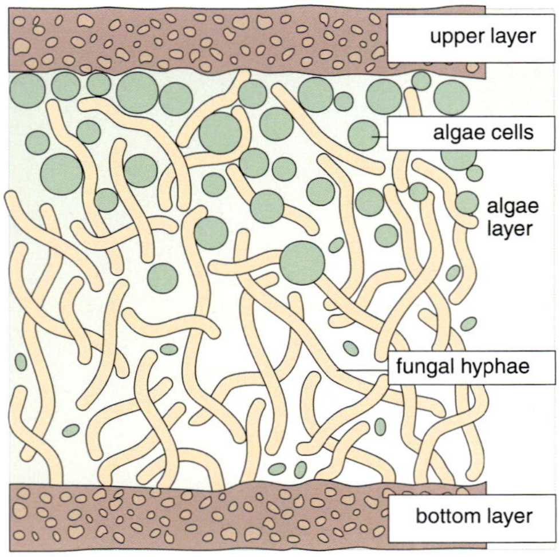


2 Lichen

## Symbiosis

Green algae (genus: *Chlorella*) can be found in *vesicles* of the protozoan *Paramecium bursaria*. These algae reproduce inside the protozoan, which provides some protection, some movement and, above all, high concentrations of minerals and carbon dioxide. The algae, in their turn, are also beneficial for *Paramecium*: if food (organic matter/*detritus*) is in short supply, the protozoans obtain vitamins and proteins from the algae. Such an interspecific relationship that is mutually beneficial is called *symbiosis*.

All species of the genus *Paramecium* sweep small food particles into their *oral groove*, where they are transferred into food vacuoles and then digested. In a similar way, *P. bursaria* can also ingest *Chlorella* algae – without digesting them however. Admittedly, if environmental conditions become less favourable and food becomes scarce, *Paramecium* starts to behave like its relatives: it digests these algae and lives on them. During evolution, these unicellular organisms seem to have gained the ability to control the release of digestive enzymes. This example shows that symbiosis does not necessarily mean a peaceful coexistence, but can often be considered as the *balancing of forces*. Only some species of *Chlorella* manage to live inside *P. bursaria*, others being digested immediately; this indicates that this species of algae has developed some resistance against the digestive system of *P. bursaria*.

| | A impacting on B | B impacting on A |
|---|---|---|
| parasitism | ⊕ | ⊖ |
| predator/prey | ⊕ | ⊖ |
| symbiosis | ⊕ | ⊕ |
| parabiosis | ⊕ | 0 |

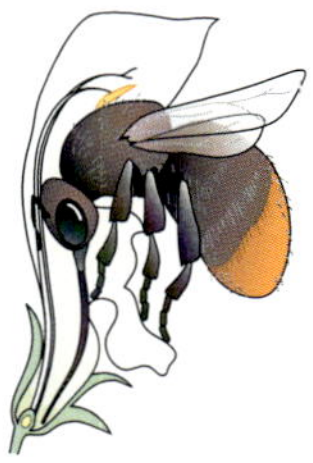

flower adapted to pollination by bumblebees

flower adapted to pollination by hummingbirds

Symbioses in nature are frequent, e.g. *lichens*. The relationship between algae or *cyanobacteria* and the surrounding fungus is so close that lichens are classified in an independent systematic group. Lichens live under entirely different conditions from those of the two constituent organisms of this symbiosis. Lichens not only populate extremely infertile ground such as rock or concrete, but also living spaces exposed to extreme conditions, such as drought or cold, where neither fungi nor algae could live on their own. The structure of the fungus prevents the algae from drying out and supplies carbon dioxide. It can provide important minerals to the algae by dissolving them from the rocky ground by giving off acids. In return, the fungus, which is a *heterotrophic* organism, depends on the organic products of photosynthesis as made by the algae. On the one hand, the coexistence of two species in a symbiosis allows the population to colonise extreme habitats; on the other hand, it means that these symbionts are stenopotent, especially with regard to certain ecological factors, as the requirements of both symbiotic species have to be met. Many types of lichen are extremely sensitive towards the pollution of the air ($SO_2$) or of the soil (pesticides) and can be used as *indicator organisms*.

There are also symbioses between species which are locally more independent of each other. Flowering plants and the insects or birds pollinating them are a good example.

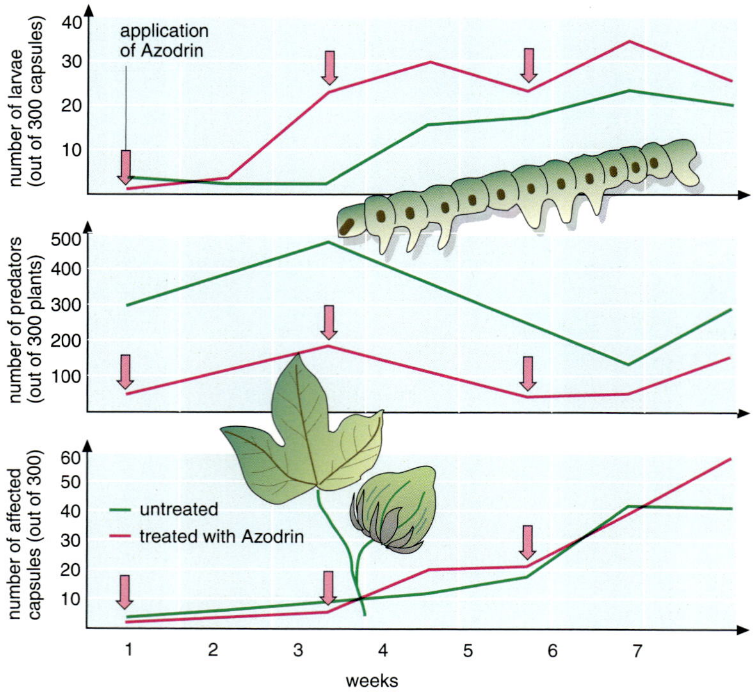

**1** Caterpillar of the cotton worms on cotton

growth, the predator population recovers more slowly than the pest populations after the application of insecticides (see Third LOTKA-VOLTERRA equation p. 27).

Some pests are almost impossible to control by insecticides. Larvae of the *cotton worm* live in the stems and capsules of cotton plants (*leaf miners*) and can significantly affect the yield of a crop. Trying to control them by externally sprayed-on insecticides is difficult as little of the insecticide reaches the pests. *Predatory* insects and spiders preying on the cotton worm however live on the surface of the plants and therefore are directly affected (fig. 1).

Plants can be genetically modified to produce insecticides by themselves: e. g. Bt-corn. It contains genes of a bacterium (*Bacillus thuringiensis*). These enable the plants to synthesise toxins that do not affect vertebrates, preying insects or spiders, but that are lethal for herbivorous insects. The use of *transgenic* plants is however not unequivocal.

The frequent application of the same substances can lead to the survival and unrestricted proliferation of any individuals that happen to be *resistant* to these substances, which then lose their effect. This is especially true for transgenic plants as they permanently produce the same insecticide.

Parasitic fungi can also reduce crop yields; *fungicides* are used against them.

## Population ecology and pest management

The growth of plant populations such as corn or vegetables, which are important as food crops, is strongly limited by herbivores ("pests") and competing plants ("weeds").

### Insecticides against pests

For herbivores, monocultures offer ideal living conditions and allow high reproduction rates. Compared with mice, populations of insects grow more quickly and are normally more mobile, which is why their impact on crops is extremely serious. The potential toxicity of *insecticides* (chemical pest control) or the products of their decomposition can make their use problematic (see p. 63). Furthermore, it does not always show the desired effect of protecting plants, as these substances interfere with the food web at various positions.

**Leaf miners**
Herbivores whose larvae live inside plants and "mine" tunnels in the plant tissue

As most predators of pest insects are also themselves *arthropods* (e. g. spiders), their populations might concomitantly be reduced if the insecticide is not sufficiently *pest-specific*. Because of their lower rate of growth, the predator population recovers more slowly than the pest populations after the application of insecticides.

### Herbicides against competitors

Arable land offers ideal living conditions to both the crop plants and to fast-growing plants such as hemp nettle, knotweed and other "weeds" that compete for limited resources such as space, light and minerals. In order to avoid smaller crop yields, *herbicides* are applied. However, not only the competing plants, but also useful insects disappear as their larvae often depend on these types of "weed" as their food source.

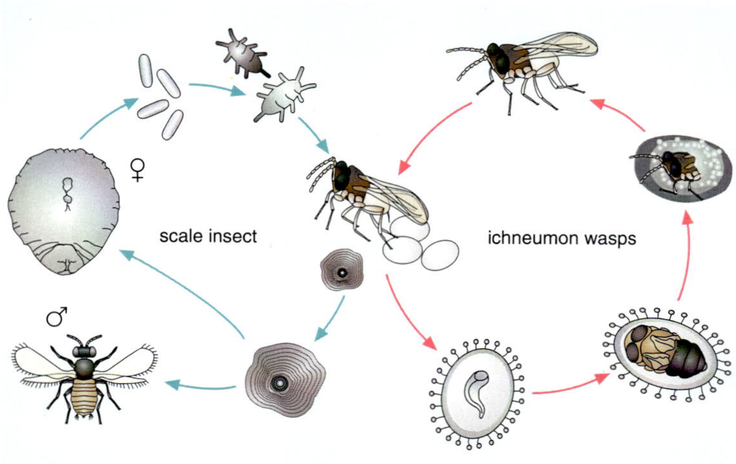


**1** Ichneumon wasps as pest control

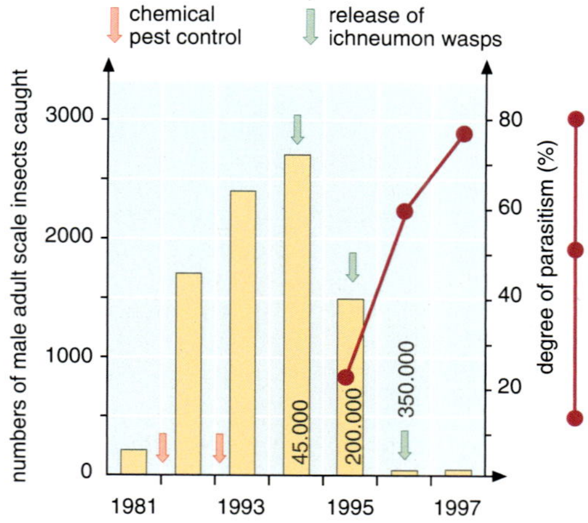


**2** Chemical and biological pest control

## Parasites and pest management

Specific *chemical pest control* has only been possible for about 60 years. Before that, people tried to combat pests with simple means. Apart from simply collecting pest insects by hand, they attempted to promote predator populations. *Hedgerows* are not only useful for wind protection, but also offer nesting locations for birds feeding on insects. *Fallow land* was once important as an area of retreat and development for predator insects and ground-nesting birds. Such ecological means of pest control have been revived and are of great significance for *organic farming*.

caterpillar *(Lymantria dispar)*, a destructive pest of shade, fruit and ornamental trees

The direct introduction of predators against pests has a long tradition. It was quickly understood that the species introduced must have some specific features so that they are both effective and do not have negative effects on the ecosystem themselves. Besides a high rate of growth they must have a high *rate of feeding*. It is also crucial that they are *prey-specific*: generalists have frequently been found to have a disastrous effect on ecosystems. Domesticated cats, originally introduced on tropical islands as a pest control against mice, have become feral and have exterminated numerous local animal species.

*Parasites* can fulfil these criteria with their high rates of growth and their being *host-specific*. Ichneumon wasps for example are used as a pest control against scale insects in orchards (fig. 1, 2). As neither predators nor parasites completely exterminate their prey/hosts, but simply keep their numbers in check, some reduction in yields will still be observed. Parasites often have to be reintroduced at regular intervals as their populations are not constant because of low host density; this can be quite costly.

The introduction of e. g. bacteria and viruses has been successful in the control of forest pests. The development of resistance among these organisms must however be taken into account. In 1950, the *Myxoma virus* was introduced into Australia to reduce the large number of rabbits there. In the following decades, 95 % of all rabbits died. In the meantime, half of the rabbit population has become resistant against the virus and their numbers are growing once again.

*Integrated pest management* is a finely tuned combination of various biological and chemical methods. Pesticides (insecticides, fungicides and herbicides) are only applied if pest numbers exceed defined thresholds despite the use of biological measures.

### Tasks

① Analyse the consequences of the application of Azodrin against cotton worms (fig. 32. 1)

② Compare the effects of chemical and biological pest control against scale insects by using the example of a fruit growing area in Switzerland (fig. 2).

## System levels

**Biocenosis**
All organisms coexisting in a biotope

**Biotope**
Environment (living space), i.e. the sum of all abiotic environmental factors

**Ecosystem**
Biocenosis and biotope together

**Biomass**
Entire mass of all living individuals

**Production**
New biomass produced over a certain time

Many species of plants and animals coexist in an ecosystem. Together they form a *biocenosis* living in their common environment, i. e. their *biotope*. The biocenosis and biotope together form an ecosystem.

The sum of the numerous interactions between populations in an ecosystem defines its rules. Some of these can be observed in the relationship between two populations. The entire system can be imagined as the sum of single systems between two species. However, a great variety of relationships exist between the species involved and therefore this must be seen as an intricate network. The approach needed to analyse a biocenosis must reflect this complexity.

### Trophic levels

Many populations make use of similar resources and are therefore summarised in *trophic levels*. These can be used to determine simple governing rules. On the bottom levels we find *autotrophic* organisms such as plants. They produce energy-rich organic compounds from inorganic substances (*primary production*). This is the trophic level of the *producers*. On all other higher trophic levels we find *consumers*. These are *heterotrophic* organisms obtaining their biomass from food, i. e. by feeding on the biomass of other organisms (*secondary production*).

The second level is occupied by herbivores, which are *primary consumers*. Carnivores are *secondary consumers* and occupy the third level. Higher levels of consumers might also be present, but with every higher level food becomes rarer and changes have more serious effects. Most ecosystems therefore do not consist of more than four levels.

Trophic levels are assigned characteristic data, i. e. numbers of *individuals/density*, *biomass*, and *production*. Because of large differences in mass and size between different organisms (e. g. trees vs low vegetation), the number of individuals has little relevance. Biomass comprises the mass of all individuals together. Production defines how much additional biomass is produced over a certain time. These figures can be displayed in *pyramids of numbers*, e. g. for biomass (fig. 1). The individual's need for space can also be displayed in a pyramid; here the proportions are reversed as, for example, a mouse needs smaller hunting grounds than an owl, which lies on a higher trophic level. The need for space usually increases with every trophic level.

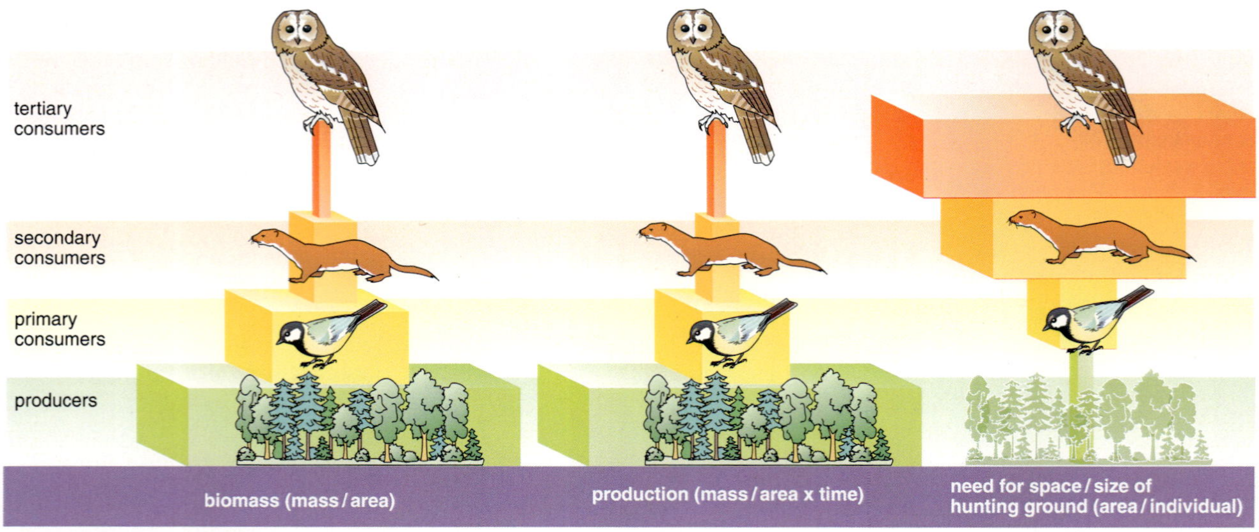


**1** Ecological pyramids of terrestrial ecosystems

| Species | Mass (kg) | Density ($N/km^2$) | Mass / area ($kg/km^2$) | Nutrition |
|---|---|---|---|---|
| Field mouse | 0.02 | 3300 | 66 | H |
| Field vole | 0.032 | 2500 | 80 | H |
| European bison | 900 | 0.1 | 90 | H |
| Boar | 80 | 0,3 | 24 | O |
| Badger | 12 | 1.2 | 14.4 | O |
| Weasel | 0.12 | 10 | 1.2 | C |
| Lynx | 25 | 0.03 | 0.75 | C |
| Wolf | 40 | 0.02 | 0.8 | C |
| H = herbivore, C = carnivore, O = omnivore | | | | |

**1** Ecological data of various mammals

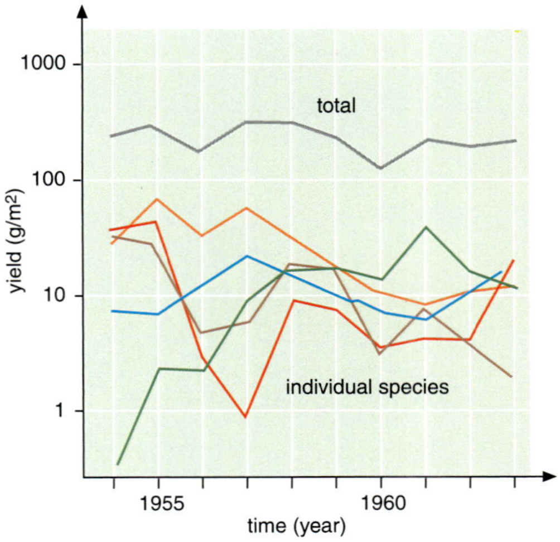


**2** Production of plant species on a meadow

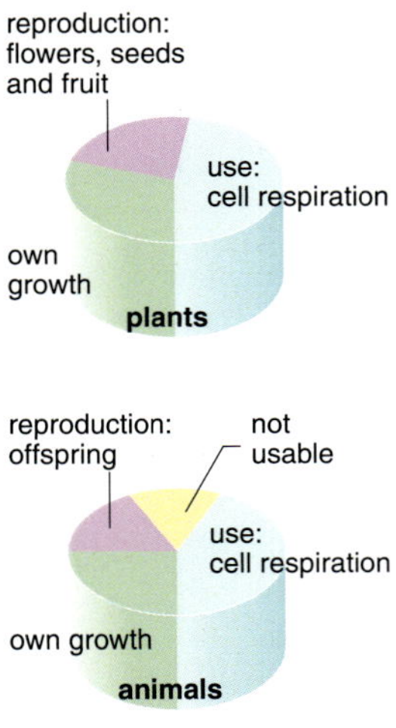


use of biomass

The size of a population does not stay the same but is continuously changing. Nonetheless, in environments with stable living conditions, the dimensions of the trophic levels seem to stay the same over long periods of time (fig. 2). In these cases, the loss of biomass through predation or deaths from each trophic level is compensated by the new production of biomass.

## Productivity ratios

Production is often smaller than the actual population. In this case, populations on lower trophic levels are larger than those on higher levels. Hence, over the same area, herbivores can attain more biomass than carnivores (fig. 1).

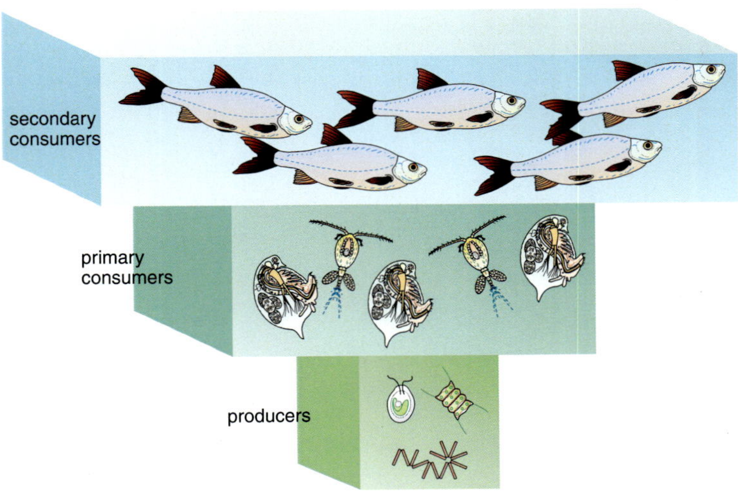


**3** Pyramid of numbers displaying biomass in a large body of water

Biomass eaten by consumers cannot be digested entirely. Most of the actually useful parts go into energy supply (cell respiration) and are also needed for individual growth and the production of offspring (*reproduction*). Together they represent the production of a trophic level.

The production of a higher trophic level is therefore lower than the amount of biomass taken in; this is why on higher trophic levels the production is always lower than that on the level below and keeps on decreasing with every level upwards. Research shows that the production of any trophic level is about 1/10 of the production of the trophic level below. The ratio of production between subsequent levels is therefore almost 10 : 1; in other words, the *efficiency of production* is 10 %.

In large bodies of water, a similar but reversed situation can be encountered; the pyramids of numbers for biomass seem to be upside down (fig. 3). If the supply with resources is good, small algae can divide several times a day and multiply their biomass; they have a high rate of production. For any ecosystem the actual biomass is therefore of lower significance than the production, which is crucial and which is always higher at lower trophic levels. Only the pyramid of numbers displaying production always has the expected pyramidal shape.

**1** Light utilisation in forests and meadows

Green parts of plants absorb light for photosynthesis. However, light cannot be absorbed entirely so that the remaining light can reach other layers of leaves. Research has shown that there are so many leaves in a stock of plants that virtually all the light is used (fig. 1).

If production is only limited by the amount of light available, different ecosystems have similar rates of photosynthesis and therefore similar net primary production. *Deciduous* woodlands, *coniferous* forests and meadows of central Europe are a good example.

Primary production is furthermore determined by temperature and the availability of water and minerals. The factor that is least available limits the overall potential yield (*Liebig's law of the minimum*, see p. 14); this results in a highly different primary production in the various environments found on Earth (fig. 2).

## Primary production

**Net primary production/year**

deciduous wood
1.45 kg / m² · a

coniferous forest
1.58 kg / m² · a

grassland
1.42 kg / m² · a

arable farmland
approx. 1.1 kg / m² · a

Plants are at the start of many food chains as they produce their own energy-rich biomass through photosynthesis. The entire production of biomass by plants is named the *gross primary production* (*GPP*). Most of it is used up by the plant in cell respiration; the rest is called the *net primary production* (*NPP*). Plants do not feed on other organisms as they are autotrophic. All other heterotrophic organisms feed, directly or indirectly, on the biomass produced by plants. The extent of primary production determines the amount of resources available in any ecosystem and therefore also the numbers of consumers that can be part of it.

### Tasks

① Regions where mineral-rich water rises from the seabed are found along coastal areas of the oceans. Give reasons why production in these regions is especially high.

② Primary production varies greatly in different parts of the world (fig. 2). For each case, explain which factor is probably limiting.

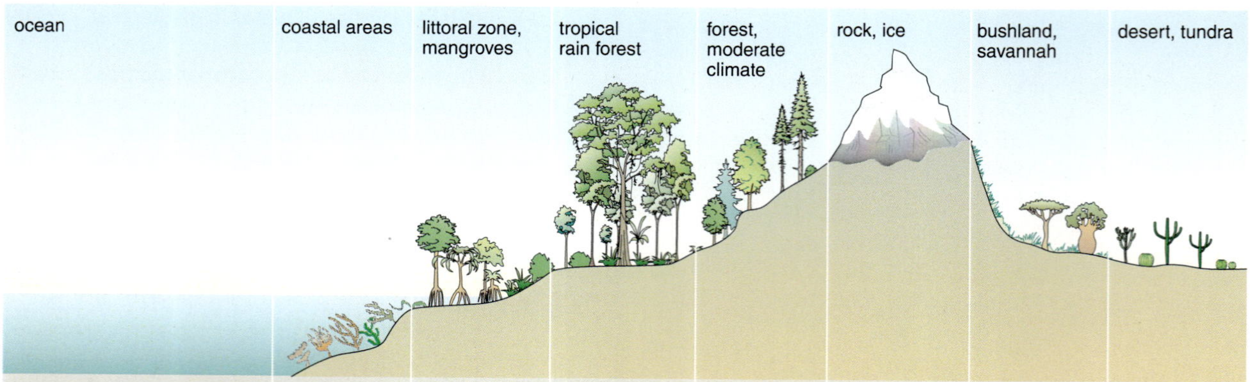

| | ocean | coastal areas | littoral zone, mangroves | tropical rain forest | forest, moderate climate | rock, ice | bushland, savannah | desert, tundra |
|---|---|---|---|---|---|---|---|---|
| net primary production (kg/m² x a) | < 0.4 | 0.2 – 0.6 | 1 – 6 | 1 – 3.5 | 0.5 – 2.5 | < 0.001 | 0.2 – 2.5 | < 0.04 |
| biomass (kg/m²) | < 0.01 | 0.01 – 0.1 | 10 – 50 | 40 – 80 | 10 – 50 | < 0.3 | 2 – 15 | 0 – 2 |

**2** Biomass and net primary production

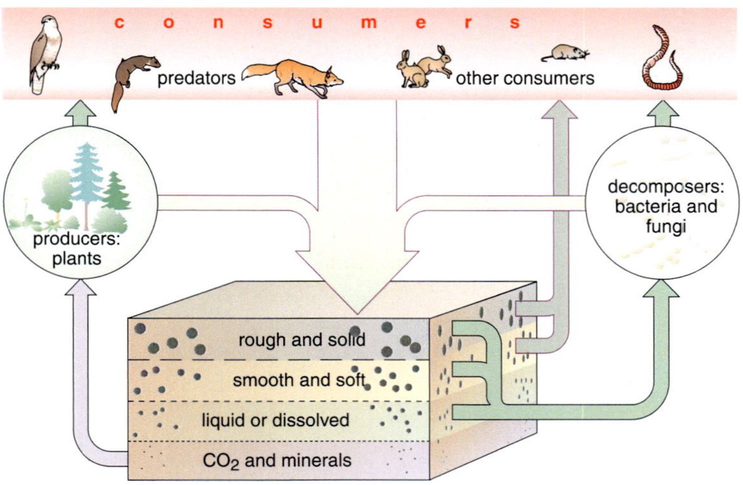

**1** Decomposition of dead organic matter

decomposition of a leaf

## Decomposition of matter by decomposers

Ecosystems contain large amounts of dead organic matter such as faeces, corpses of dead organisms or debris, including the leaves and flowers of plants or the feathers, hair and skin particles of animals. As such organic matter represents energy, many organisms (*decomposers*) use it as a resource. Unlike predators that kill their prey and impact on their growth and reproduction rates, decomposers depend on resources that they find as dead organic matter.

Many organisms are part of the decomposition process, among them bacteria and fungi. In water, dead organic material (*detritus*) is used e. g. by filter feeders. In forests, mites, small insects, woodlice, worms, snails and other small animals feed on larger pieces of organic matter.

Whether a particle of matter is useful depends on its size. Large pieces such as carrion are eaten and digested, i. e. broken down into liquids or water soluble substances by enzymatic reactions and can then be reabsorbed. Whatever cannot be used in this way is excreted as faeces. Therefore, decomposers themselves leave behind more dead organic matter that can be used by other decomposers. Every time the material is used, its amount, particle size, and energy content are decreased.

Chemical composition is another factor that determines the usability of dead organic matter. Liquid or soluble substances can be used directly, which is why proteins and carbohydrates (apart from cellulose) are decomposed quickly. Every time things are used, the remainder consists of substances that are harder to decompose, e. g. chitin, cellulose or lignin, which can only slowly be decomposed by specialists such as certain species of bacteria and fungi.

Decomposition processes need water as corpses otherwise dry out and mummify. High temperatures and a neutral pH value speed up the decomposition process. In central Europe, for example, leaf litter can take a number of years to decompose completely (mineralisation), although under better conditions, the time taken can be reduced to only a year.

Dead organic matter is decomposed until all useable energy is gone, which is the case once only carbon dioxide and minerals are left. Originally, primary producers used these very substances to produce biomass. Through decomposition they are released and can again be used by producers. Carbon dioxide and minerals are not used up, but go through cycles of fixation and release. Decomposers remineralise the biomass remains and so close this cycle; herein lies their great importance for any ecosystem.

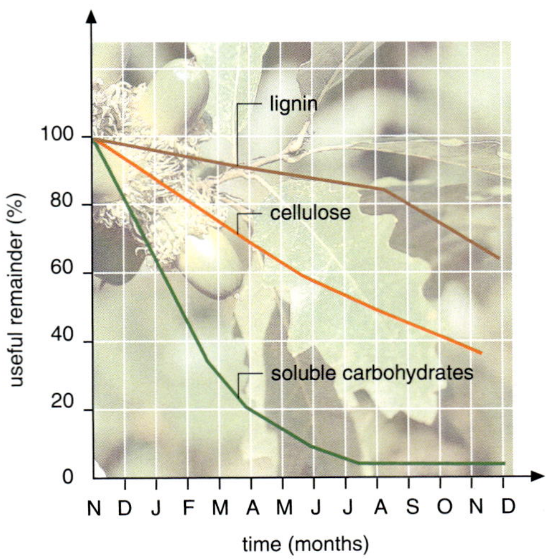

**2** Decomposition of oak leaves

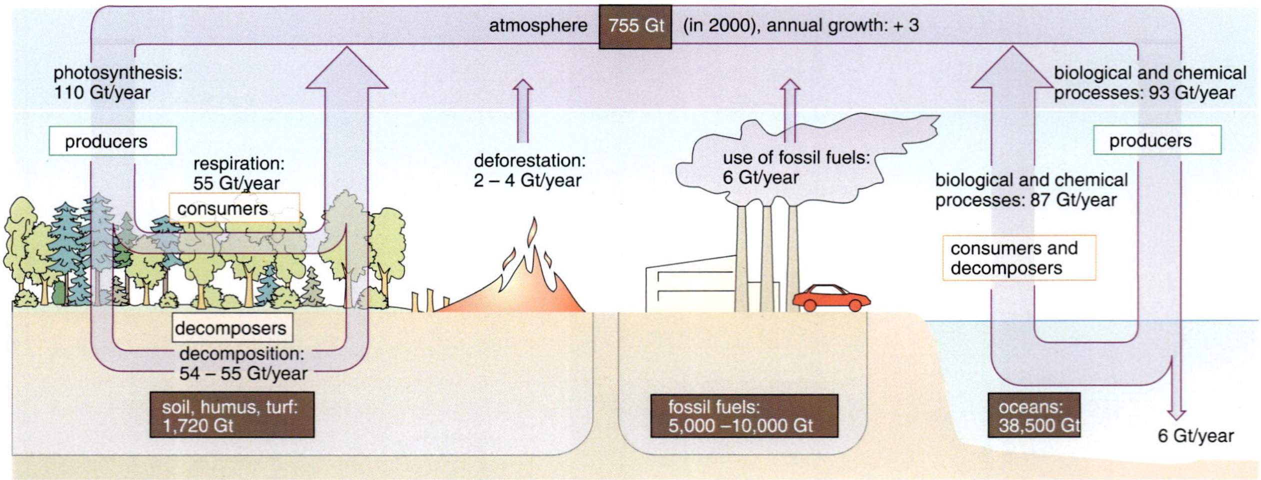


**1** Global carbon cycle (all data in gigatons/year; Gt = gigaton)

## Cycles of matter and energy flow

Inorganic substances are fixed as biomass by primary producers and are re-released by consumers and decomposers. Chemical elements such as carbon or nitrogen therefore pass through theses cycles.

### Carbon cycle

*Carbon* is the most important element in biomass. All biomass consists of carbon compounds that are made by producers using carbon dioxide ($CO_2$). These carbon compounds are employed as both structural compounds and as energy sources. Through feeding and decomposition they are passed to consumers and decomposers respectively, which use them for either building their own biomass or for energy through respiration; this reforms inorganic carbon dioxide. These processes continue until all carbon originally fixed as biomass is re-released as carbon dioxide, which closes the carbon cycle.

Plants on land and in the oceans fix part of the carbon present in the atmosphere. Another part is fixed through processes that change $CO_2$ into carbonate (mainly calcium carbonate, i.e. limestone). Sooner or later, all the carbon returns as $CO_2$ into the atmosphere; hence, statistically all the $CO_2$ is turned over every three to four years within the biosphere.

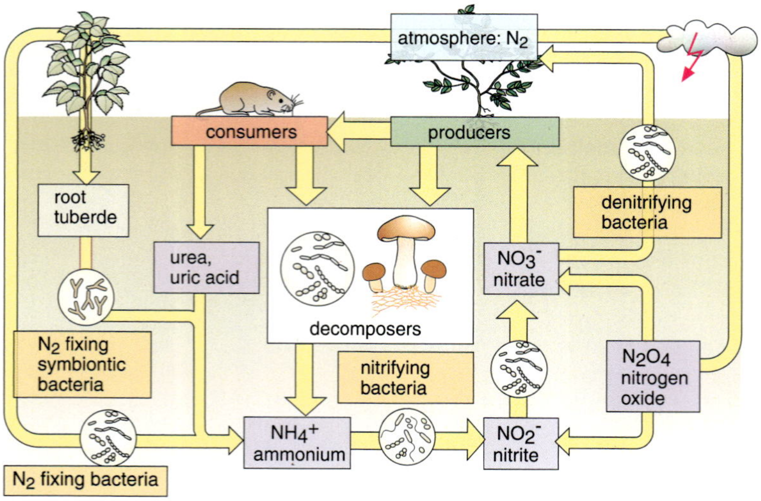


**2** Nitrogen cycle

### Nitrogen cycle

*Nitrogen* is an essential element of biological molecules, e. g. amino acids or nucleotides. Plants take in nitrogen atoms as nitrates ($NO_3^-$) or ammonium ($NH_4^+$) from the soil and integrate them into their biological molecules: annually about 175 million tons of nitrogen are fixed by plants. Some bacteria are even able to use the chemically more stable atmospheric nitrogen ($N_2$); some of these bacteria live as symbionts in the *root tubercles* of specific plants. Consumers and decomposers put these nitrogen compounds to their own use.

Excess nitrogen is excreted as ammonia or urea/uric acid. The latter are changed in the soil into ammonium; via the intermediate compound nitrite ($NO_2^-$), ammonium is turned back into nitrate by nitrifying bacteria and again becomes available for plants. Denitrifying bacteria change nitrates into $N_2$, closing the whole cycle by returning $N_2$ into the atmosphere.

All organisms use part of their food intake for their metabolism, which releases energy as heat. In the end, all of the sun's energy is transferred into heat energy. The flow of energy through the organisms' biomass therefore only represents a delay that enables life on Earth.

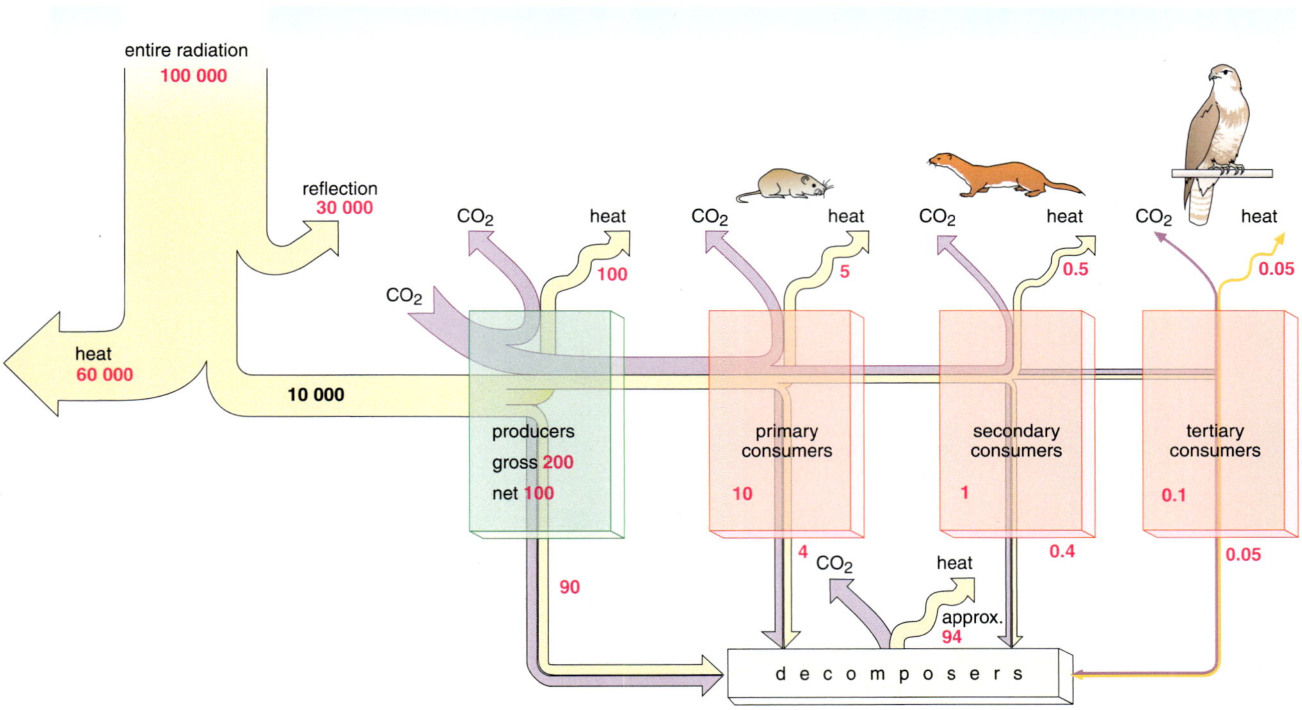


**1** Daily flow of energy (kJ / $m^2$ · day)

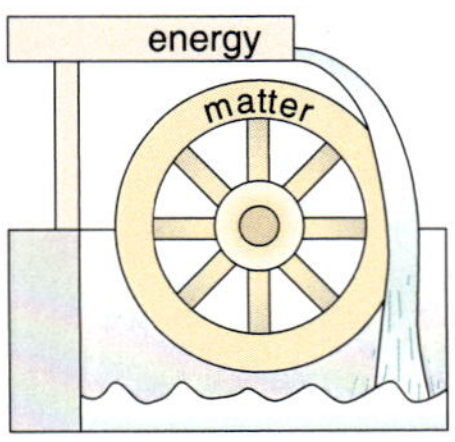


## Flow of energy

The cycles of carbon, nitrogen and other elements are powered by the energy of the sun. Producers store this energy as chemical energy in organic compounds. The energy is transferred to consumers and decomposers as organic compounds. This shows the close link between energy and biomass. Only 10 000 kJ/$m^2$ of all the sun's energy hitting the Earth each day actually reaches the Earth's surface. Only 2 % of this energy is used for photosynthesis. Half of the products are used for cell respiration, whereas the other half is stored in plant biomass.

## Tasks

① Use figure 38. 1 to contrast processes that release $CO_2$ (sources) with processes fixing $CO_2$.

② Calculate, from sun's daily radiation hitting the atmosphere, the percentage that reaches the Earth's surface.

③ A water mill is used as a model to show the close link between the cycle of matter and the flow of energy. Explain the advantages and limitations of this model.

# Structures of ecosystems

Ecosystems are not homogeneous but are *spatially* structured; these structures are structurally subdivided themselves. Furthermore, as structures do not stay the same but undergo change, a time dimension is also involved.

## Spatial structures

ecosystems are like boxes

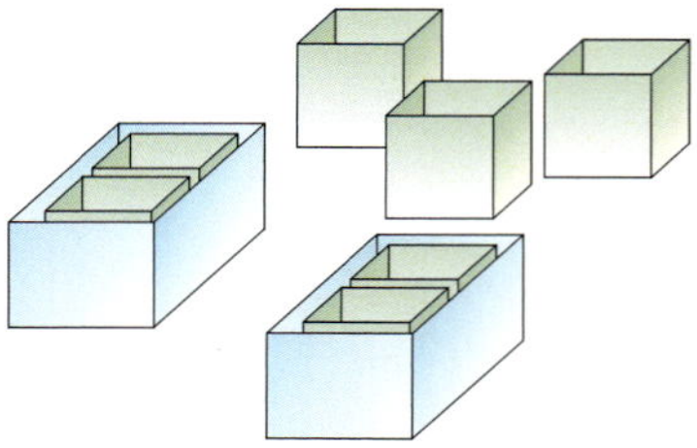

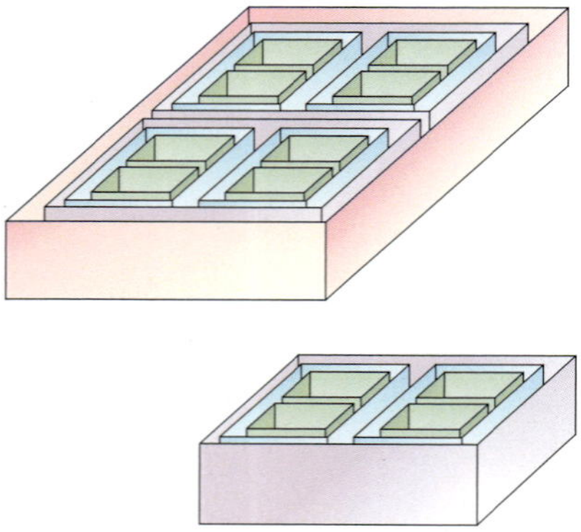

ecosystems are like islands

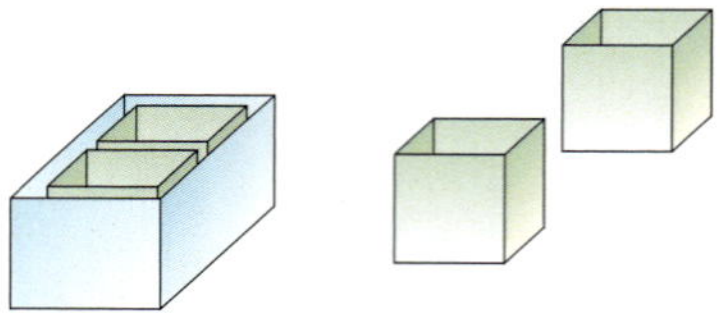

**Explain why both the box and island analogy show aspects of the same ecosystem. Use an example to show the characteristic box and island features.**

## Stay or go?

"*Expansion* is a basic feature of organic nature itself. The life cycle of any species goes through phases of expansion. The purpose of expansion is to allow the species to populate unoccupied areas or only partly filled niches."
(E. MAYR)

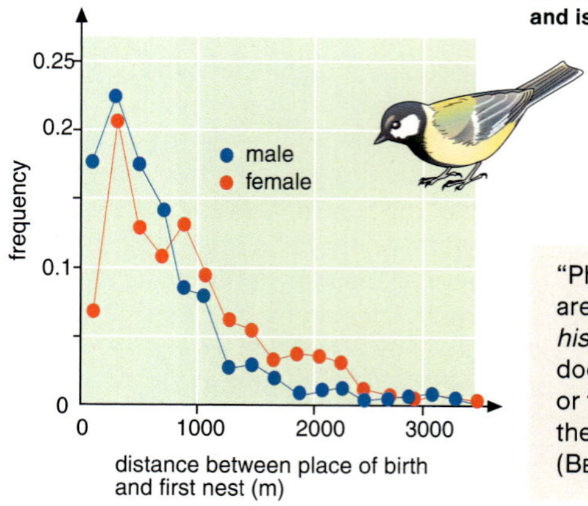


"Phases of expansion and migration are of crucial importance in the *life history* of all organisms; they are doomed to become *extinct* if they or their offspring always remain in the same area".
(BEGON, HARPER & TOWNSEND)

## Temporal structures

Organisms are usually adapted to their environment.

**Choose an organism and explain how it can respond to changes in its environment.**

Cycle of mussel banks

**Compare characteristic local features of an ecosystem with its temporal features. Can the individual phases of a system be compared with "islands of time"?**

**Find information about the life cycle of an organism and indicate when expansion is possible. How can organisms in a lake, for example, manage to populate new lakes?**

plant on volcanic rock

## Ecosystems are transient

For the *boreal forests* of Alaska, a natural cycle of fire is postulated that normally takes between 50 and 200 years, depending on the prevalent tree species and region. Fire has the ecological effect of removing layers of *duff*, i.e. humus and leaf litter that accumulates over long periods of time and that can impede the natural regeneration of the tree population. It also immediately releases all the minerals contained in these layers of organic material and makes them available to plants. In the long run, trees would suffer from a lack of minerals if it was not for these fires.

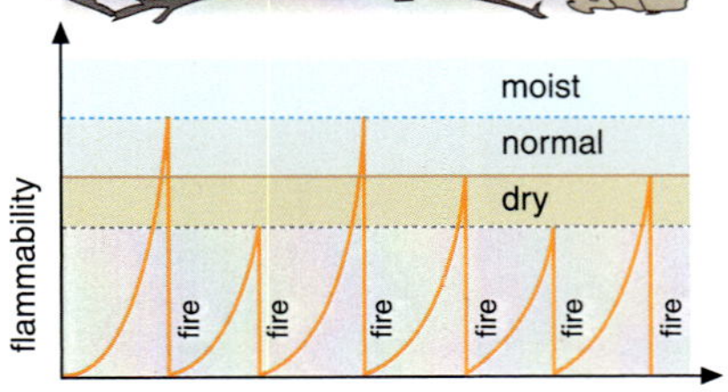


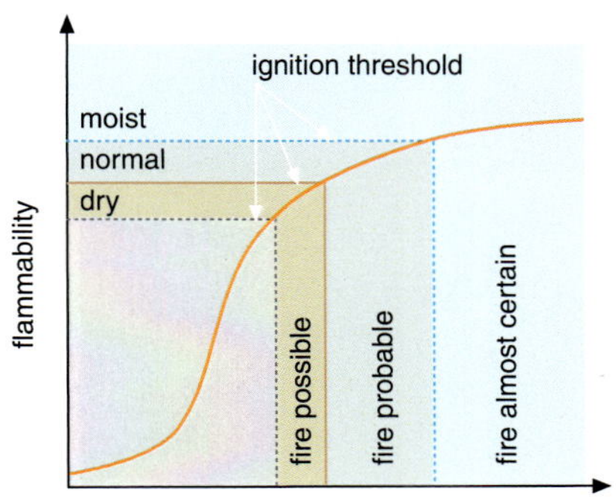

# Ecological successions

In places such as volcanic islands or fallow fields, new and uninhabited living spaces come into existence and are quickly populated by immigrating organisms. Organisms often appear in a typical order when they start building biocenoses. Such a sequence is called an *ecological succession*.

## Phases of succession in a forest

On clear-felled areas in forests, for example, a highly typical succession of plant communities can be observed. This is mainly caused by the plant species competing for light, i.e. their production rates through photosynthesis (*autotrophic succession*).

The process can be subdivided into three phases. During the *initial phase* (primary succession), the area is densely populated by *pioneer plants* — often only a few different species. Usually, these are small annual plants that are not very competitive, such as grasses, i.e. typical r-strategists (see p. 28). They are then displaced by slightly taller growing and therefore more competitive perennial plant species such as small shrubs. In the second phase (secondary succession), these plants are in turn displaced by larger species that cast shadows over the smaller plants and so outperform them. A large amount of plant biomass is produced during this phase. After many years, the largest tree species dominate as their size makes them the most competitive. Both the initial and second phase are highly productive phases of the succession (fig. 2). In the final phase (*climax* community), *dynamic equilibrium* occurs between the production and decomposition of biomass. The number of species decreases, but the species surviving are usually long lived.

Animals also show an ordered succession. They populate areas according to the availability of the resources they need.

Many decades are required to reach the climax stage. The structure of this stage mainly depends on the overall climatic conditions; this explains the strong variation seen in different parts of the world.

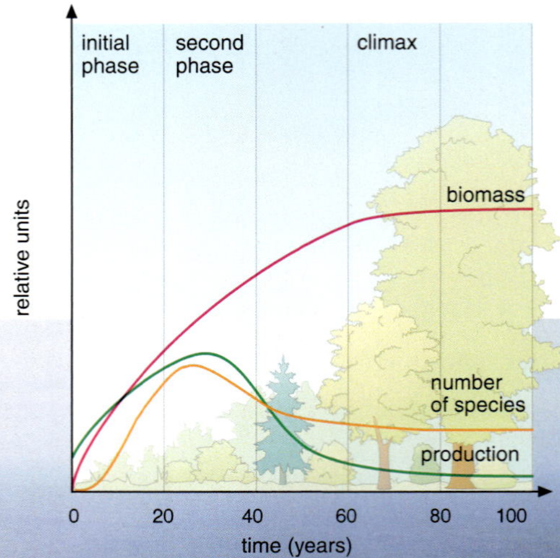


**2** Succession phases in a forest

**1** Volcanic island of Surtsey (formed in 1963)

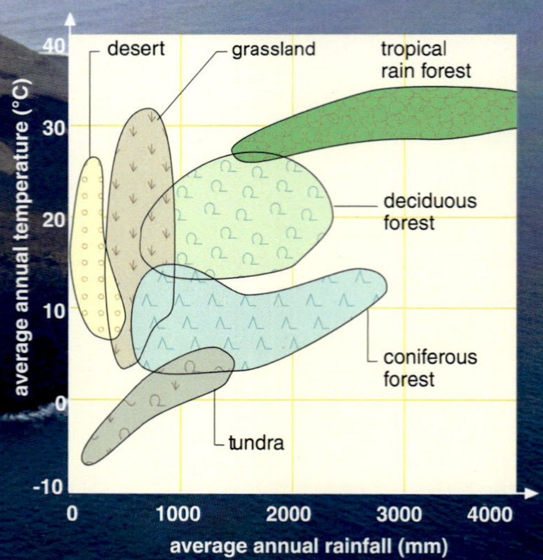


**3** Climax communities of the world

### Mosaic cycles

The sequence of succession as described above assumes homogeneously structured phases. Although this is helpful for the analysis of patterns, this view is too simplified for a detailed analysis. Disturbances in the climax forest, e. g. damage through storms or the falling of old or dead trees, create small clearings. Animals often live in groups so that their distribution is uneven. They can cause damage through feeding or by trampling the undergrowth with their feet. Locally restricted disturbances such as these set an area back into earlier phases and a new succession can start. In a long-term view, many of these local new successions can exist in close *proximity* and each of them can be in a different phase of succession. Over a large area, this has an appearance of a mosaic of smaller islands of succession (fig. 1), the phases of which can pass through cycles.

### Permanently disturbed environments

Some environments can be considered as being permanently disturbed. River banks are restructured by each event of flooding; fields are regularly ploughed over.The same is true for mud flats, which are permanently changed by the tides. Environments such as these are populated by typical r-strategists and remain in the initial phases of a succession.

**1** Mosaic of local successions in a forest

»info box«

### Succession in a preserving jar – hay infusion

When hay is covered with water and left to stand for some days (hay infusion), organisms appear in a certain order. These organisms are unicellular, such as the various species of *Paramecium*. Their sequence is determined by the progress of the decomposition of the organic matter (hay) by bacteria. Different types of bacteria are involved in this process. The unicellular organisms feed on the bacteria. The succession comes to a close once all organic matter is used up and is completely remineralised. This is an example of a *heterotrophic succession*. The actual sequence of organisms depends on the specific conditions.

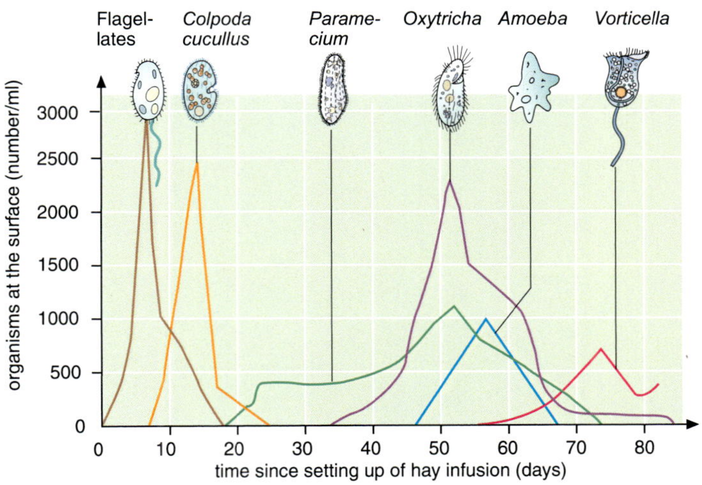

1 Forest with high / low biodiversity

## Diversity and stability

**Biodiversity**
(Variety)
- diversity of species
- genetic variety
- environmental variety

Ecosystems contain a great variety of organisms. This *biodiversity* is a distinctive feature of biocenoses. Their analysis helps us to compare ecosystems and to measure ongoing developments. For a more comprehensive understanding of the biodiversity of a certain area, various factors need to be taken into account: the numbers of all the species present, their genetic variety and the variety of ecosystems in the area. To simplify matters, only the variety of species is considered here.

The *diversity index* (see p. 48) is used as an indicator to compare the diversities of ecosystems. A large number of species being present results in a high diversity, whereas fewer species lead to a lower index. The concept of diversity also takes into account, whether certain species are dominant in an ecosystem. If one species is dominant and other species occur less frequently, then diversity is lower compared with situations where numbers are more balanced (fig. 2).

Diversity increases with the number of species and their uniform distribution.

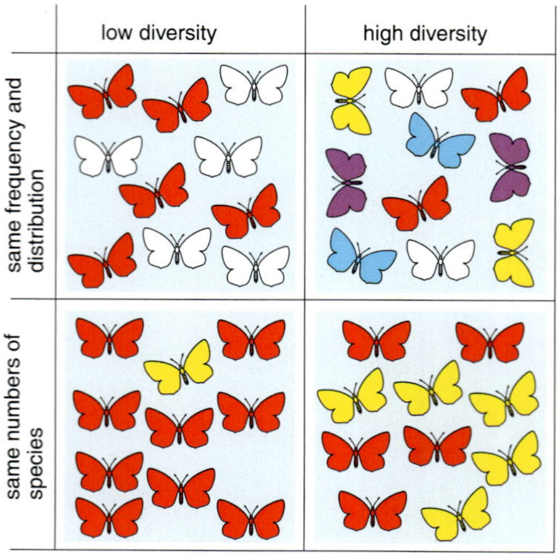


2 Aspects of biodiversity

For most environments, the actual number of species present is not known, especially for smaller species. Data concerning the frequency of species can vary considerably. Diversity calculations are thus usually carried out for comparatively small and manageable groups of organisms, e.g. plants or insects. In addition, environments are not isolated entities, but are embedded into a larger context (see p. 40). Investigations into diversity therefore require the definition of the boundaries of the section to be examined. Interpretation of the data obtained will hence only apply to the chosen groups inside the chosen area. Biodiversity indices on their own are thus of only limited significance and merely help to compare various ecosystems with each other.

Factors responsible for a high or low biodiversity are difficult to investigate; however, a number of feasible ideas have been proposed. These assume that diversity depends on the amount of resources available: the higher their supply in any ecosystem, the higher its productivity.

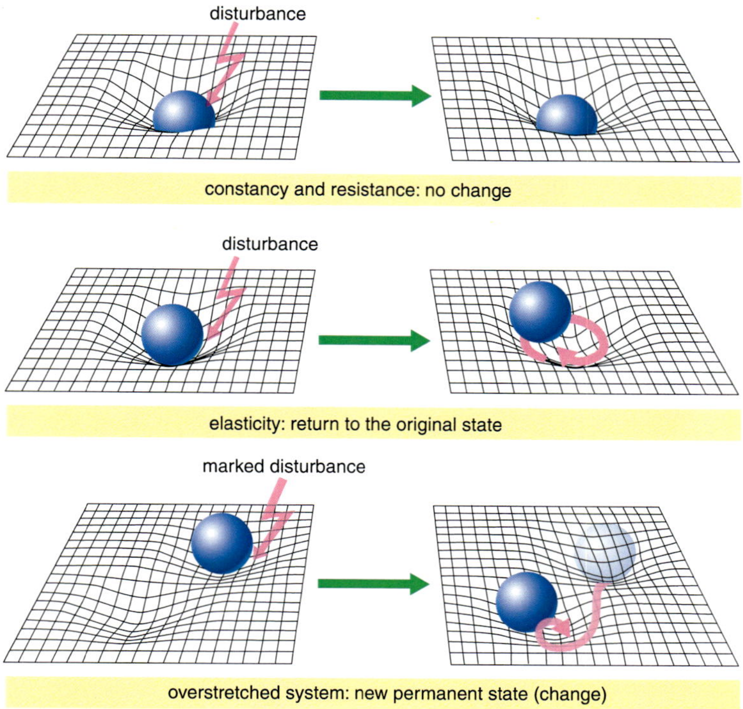


**1** Model of difference in stability

This is inevitably interlinked with the possible number of trophic levels and the number of niches. A greater supply of resources therefore also means a higher diversity. If an ecosystem can be subdivided into various structural parts, this also results in higher diversity, as a variety of structures means a greater variety of niches and consequently a greater diversity of species. Furthermore, biodiversity goes down from the tropics to the polar regions; a possible reason for this is the different climatic stability in varying geographical areas. A constant tropical climate allows, for example, greater specialisation to narrower ecological niches. More species can coexist than in temperate zones, which are under the influence of changing seasons. This is true for all ecosystems that are exposed to little or no change.

The diversity of an ecosystem often goes hand in hand with its stability (*diversity-stability-hypothesis*). Stability means that the number of species and biomass hardly change in an ecosystem, i. e. there is *constancy*. An ecosystem is also considered stable if it shows resistance towards disturbances and does not change as a result (*resistance*, fig. 1).

To explain this supposed link between diversity and stability, feasible hypotheses have been proposed. With increasing diversity, the number of possible links between species is assumed also to increase; this might protect the ecosystem from changes from external factors. Greater diversity thus accounts for higher stability.

In central Europe, forests have the greatest biodiversity in regions in which permanent small local disturbances occur through human cultivation (e. g. deforestation or the construction of footpaths), as this increases structural variety. Once cultivation stops, these forests start to return to their natural states and become increasingly dominated by the common beech, which is a strong competitor. As beeches are efficient in absorbing light, the vegetation on the ground becomes scarce. This ecological succession continues until the climax stadium is achieved, which has a smaller variety of species but is more stable than the managed forests. The assumed straightforward link between diversity and stability is therefore not appropriate.

Generalists also seem to have a stabilising effect on ecosystems. As they use a wide range of resources, they create more links between the food chains of an ecosystem than do specialists. They can cope better with disturbances that reduce their resources (see p. 29). It is not the constancy or resistance of an ecosystem that is strengthened by generalists, but its *elasticity*. This is the ability of the ecosystem to return to its original state after disturbances have caused temporary changes (i. e. to regenerate). In environments where disturbances occur frequently, the species present are adapted to these disturbances, which therefore do not have any lasting effects.

If a disturbance exceeds a certain *threshold level*, continues for a long time and is continuously acting in the same direction, the elasticity can be overstretched and can result in permanent changes to the system. This is the case with the eutrophication of lakes, i. e. when minerals are constantly added to a lake (see p. 52).

**1** Coniferous and deciduous forests of the central European landscape

## Forest ecosystems

More than 90 % of central Europe used to be covered by forest. Even though this area has decreased enormously, forests are still of great importance to us: wood is a renewable resource. People seeking recreation enjoy the cool temperatures by forests during the summer, whereas during the winter, the cold is less harsh in a forest compared with open country (fig. 2). This *attenuating* effect also occurs during the course of a day. Changes in other abiotic factors such as wind speed or air humidity are also reduced. This buffering effect creates favourable living conditions for plants and animals alike, as the impact of extreme climatic conditions is harmful to most species.

Trees are typical K-strategists (see p. 28) and indicate that forests are stable ecosystems. The high competitiveness and durability of trees makes clear, why this kind of environment is dominant in all parts of the world with a moderate climate and sufficient humidity.

Most forests in central Europe are *managed forests*. Depending on economical *feasibility* and soil composition, certain tree species are preferred for planting, some of which would not be able to compete and dominate in corresponding natural forests. A natural forest in the centre of Europe would be deciduous, green during the summer and bare during the winter. If it was not for human interference, *coniferous forests* would only occur in high mountain regions. In the northeast and north of Europe, coniferous trees are originally even dominant in flat areas, as they are more resistant towards cold and better adapted to long periods of relative darkness. The high productivity of deciduous trees during summer is counterbalanced by the ability of coniferous trees to photosynthesise all year round.

Natural deciduous forests are structured spatially and temporally (see pp. 40/41). *Layers* of low-lying vegetation, shrubs, and trees give the forest its vertical structure; it

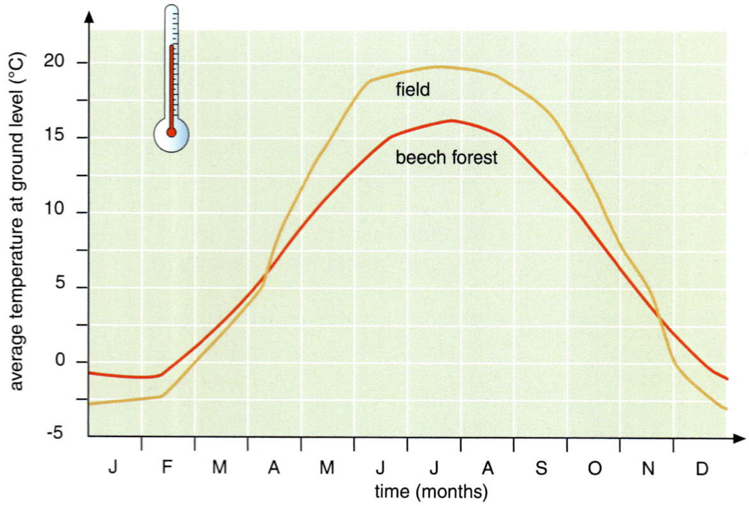


**2** Annual temperature changes

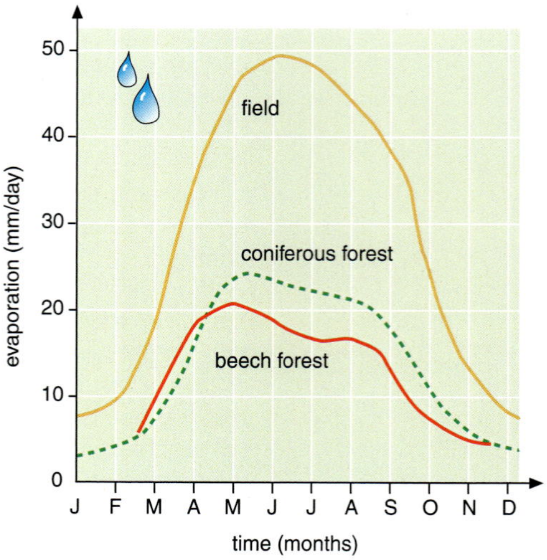


**3** Annual evaporation

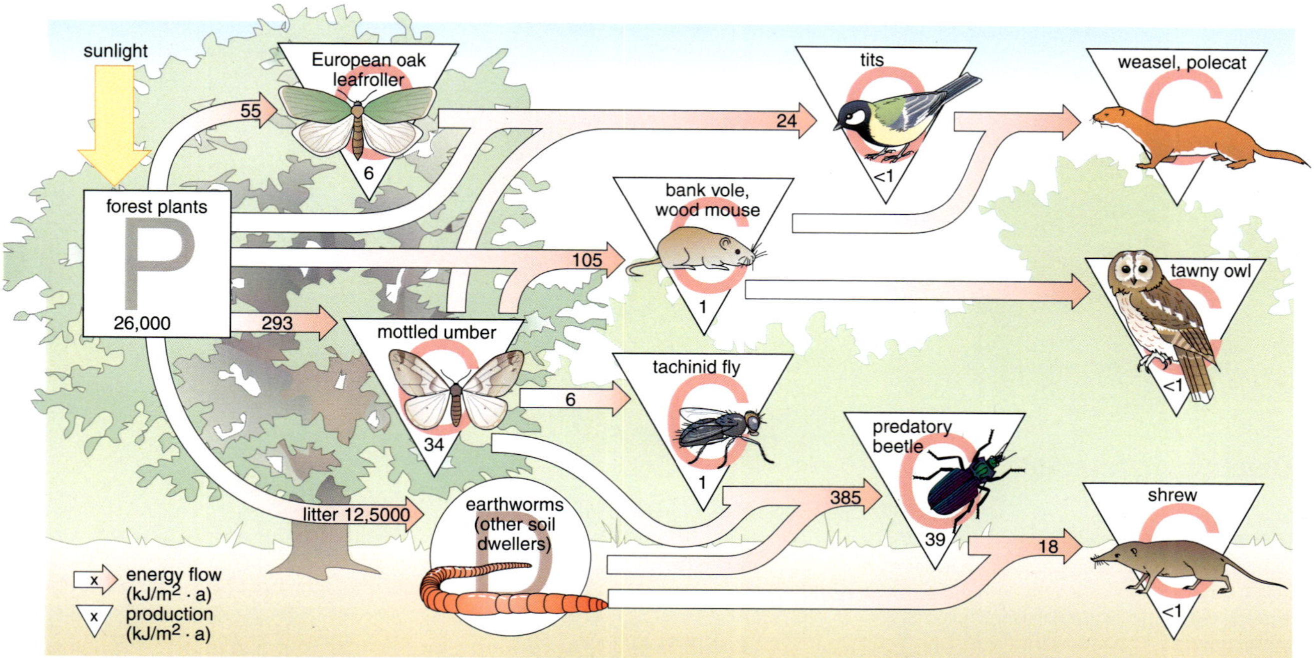


**1** Energy flow in a simplified food web of an oak forest (P = producer, C = consumer, D = decomposer)

is *stratified*. The bottom layers, especially the lower vegetation layer, consists of plants that either only photosynthesise and produce flowers during spring (early flowerers) or are adapted to low light conditions. Later during the year, first shrubs and then trees turn green. This *seasonally staged arrangement* reduces competition for light, which is a scarce resource. Early shoots however carry the risk of being damaged by late frosts.

Damage through storm, fire and local infestation with insects cause a mosaic-like distribution of tree types and ages (see p. 43) and result in a horizontal local structure. As another temporal dimension, the *age distribution* is a determinant for associated dominant plants and for fauna.

*Deadwood* is of significant importance. The substrate created by fungi decomposing deadwood is a *prerequisite* for many invertebrate species living on it. This *mosaic-like mix* of structures results in the great *biodiversity* seen in natural deciduous forests. Beech-dominated forests contain about 7000 different species of animals alone – its *diversity* is accordingly high. Finally, *natural catastrophes* also contribute to the biodiversity of forests. Some species, the North American *mammoth tree Sequoia* for example, are even especially adapted to forest fires. Their cones dry during a fire and can only thereafter release their seeds. As other competing plant species are destroyed by fire, the chances of survival for *Sequoia* saplings are particularly high.

### Taks

① "Mass infestation by herbivorous insect pests is hardly ever a real threat to forests". Give reasons for this statement.

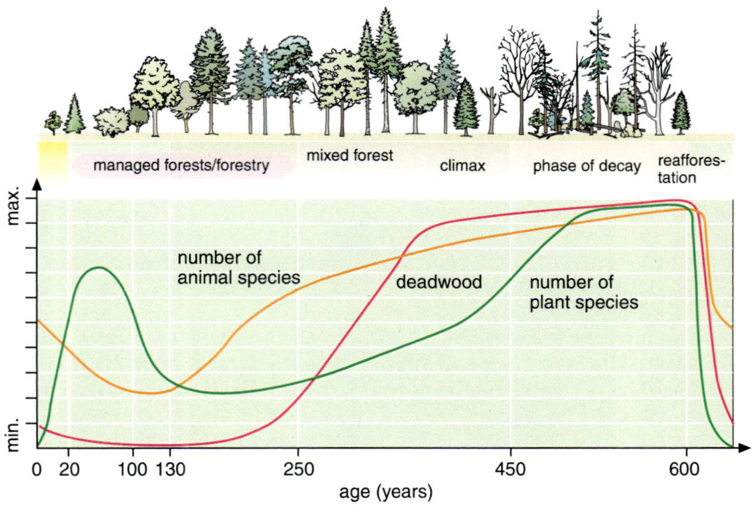


**2** Biodiversity and age structure of forests

## Diversity in forests

The determination of the diversity index is an attempt to compare the diversities of species of different ecosystems (see p. 44). Investigations regarding biodiversity can be performed in different environments. The procedure exemplified here for a forest can be adjusted for application to other ecosystems. It makes sense to compare at least two areas that show at least one distinctive difference, e.g.:

a) deciduous vs coniferous forest
b) natural forest vs managed forest (mainly monoculture)
c) dry vs humid locations
d) lawn vs meadow vs pasture vs fallow land

The areas that are investigated should be at least 50 m apart (figure top right). Additionally, abiotic factors such as light intensity (same time of day), pH value, humus content, and soil humidity could be investigated.

### Microorganisms in the soil

*Equipment:*
plastic bag, small shovel, microscope slides with a hollow grinding, cover slips, growth medium (sucrose agar or sugar solution), plasticine (or play dough), heating cabinet, microscope, stereo microscope.

#### Experiment

From each area of investigation, take about half a litre of soil (upper layers) and place it into a plastic bag using the shovel. Mix each sample.

Prepare two slides for each soil sample according to the following instructions:

- Place a fine ring of plasticine around the depression in the microscope slides.
- Place a drop of growth medium into the depression and add a small crumb of soil from the upper layers.
- Put a cover slip on top of the plasticine ring and carefully press down so as to create an enclosed space.
- Keep one slide of each soil at room temperature and incubate the other slide at 30 °C.

**Caution:** Harmful microorganisms might multiply in such samples. Please ensure that the space under the cover slip remains sealed!

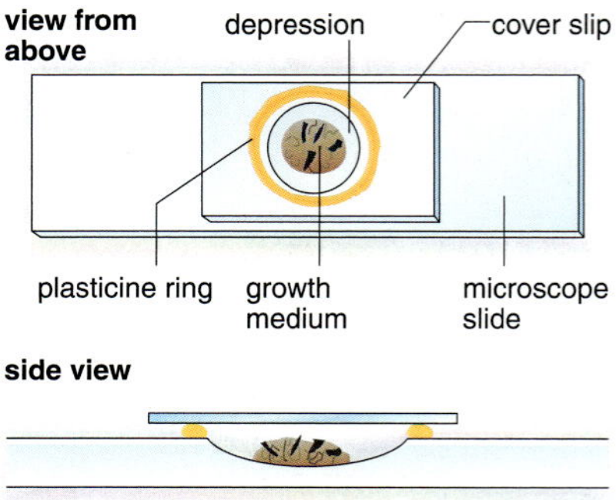


#### Task

① Using a (stereo) microscope, check the samples once every day and record any changes. Compare your observations of the different samples.

### Species-area curve

*Equipment:*
Map of area (1 : 5000), 2 m of string (with marks every 0.5 m), 4 sticks (approx. 30 cm long), labelled collection boxes (plastic bags, film canister), Petri dishes, forceps, magnifying glasses, note paper, field guides

Carry out the following procedure in each of the areas of investigation:

- Throw a wooden stick over your shoulder. It will mark the starting point of your first sample area (sample area 1). Locate and mark the area on your map.

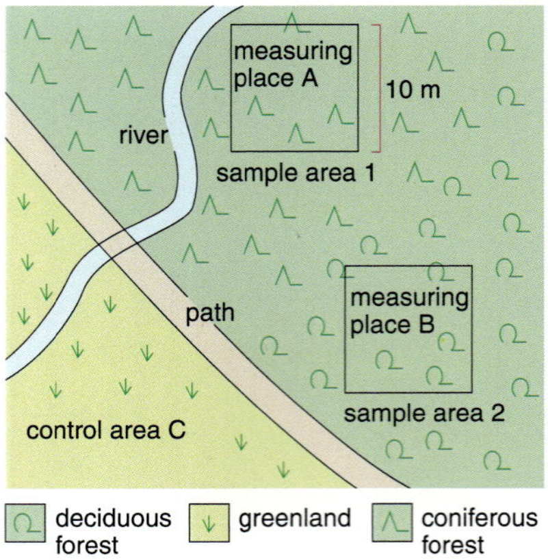


- Mark out an area of 0.5 · 0.5 m
- Document the area (sketch, photograph, verbal description).
- Identify the dominant species of plants and fungi in the sample area.
- List the number of invertebrate species in the sample area. Collect specimens of any species that you do not recognise and make notes about them. Try to identify the unknown species by using the field guides; give unidentifiable species a working title (species A, B, etc) and take notes describing their characteristic features.
- Throw the wooden stick again to obtain sample area 2 and identify the species in this area as above.
- Repeat this procedure at least ten times.

#### Complementary animal counts

**a) Counts using a Berlese funnel**

- At each location, take a sample of about 0.5 litre of soil from the surface and put it into a plastic bag.
- Place the soil samples in small portions into the Berlese funnel (see figure below) and expose to light in each case for about 30 minutes.
- Identify the number of species and numbers of individuals as above.

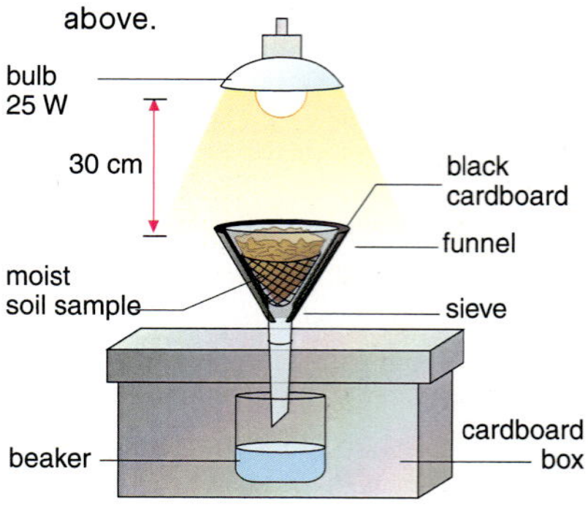

**b) Acquisition of Barber traps data**

Barber traps are a simple way of counting active and moving invertebrates; they are open containers buried in the ground, the rim lying flush with the surface.

- At each location at the diagonally opposite corners of the rectangular sample area, bury two yoghurt pots so that the openings are slightly below ground level.
- Results will be more accurate if the pots are filled 2 cm deep with water to which some washing up liquid or some methylated spirit has been added.
- In case of heavy rain it is useful to build small plastic covers over the traps without blocking access to them.
- Clearly mark the location of the traps on the map.
- Check the traps the following day and determine the number of species and individuals.
- Repeat the experiment over three consecutive days.

### Extension: plant count

Similarly, a species-area curve can be obtained for the vegetation of an area. Ten sample areas of 5 · 5 m represent a good sampling size.

#### Tasks

② Successively add up the number of species found in all samples, i. e. calculate the number of species for the areas 1 + 2, area 1 + 2 + 3, etc.
③ Draw diagrams to show the number of species in the total sample area ($m^2$).
④ Compare the different species-area curves found for the different types of forest environments. Which conclusions can be drawn?
⑤ How big should the sample area be for a representative and sound investigation?
⑥ Compare the mix of species and their numbers found with the different counting methods (searching, Berlese funnel, Barber traps). What are the reasons for the differences observed? What conclusions should be drawn for the interpretation of the data collected?

## Diversity of invertebrate fauna

This investigation can be carried out alongside that for the species-area curve (see p. 48). In this case use the counting method as described there.

*Equipment:*
as for species-area curve

### Experiment

- Collect all the organisms that you can find in the sample area of 3 · 3 m. For very frequent animals or those that are difficult to collect (e. g. ants, spiders) only collect a few specimens and note down an estimate of how many there are.
- Wherever possible identify the most prevalent species. Label any unidentifiable species with working titles (species A) and note down some of their distinguishing features.
- Record the numbers of individuals for each species (count the numbers of collected animals or, if necessary, give an estimate of the numbers in the area).

**a) Rank-frequency curve**

Example of a ranking list

| Species | Ranking | Number | Relative frequency |
|---|---|---|---|
| C | 1 | 87 | 0.73 |
| B | 2 | 12 | 0.10 |
| A | 3 | 8 | 0.07 |
| D | 4 | 5 | 0.04 |
| E | 5 | 3 | 0.03 |
| N | 6 | 2 | 0.02 |
| G | 7 | 1 | 0.01 |
| F | 8 | 1 | 0.01 |
| Total | | 119 | 1.01 |

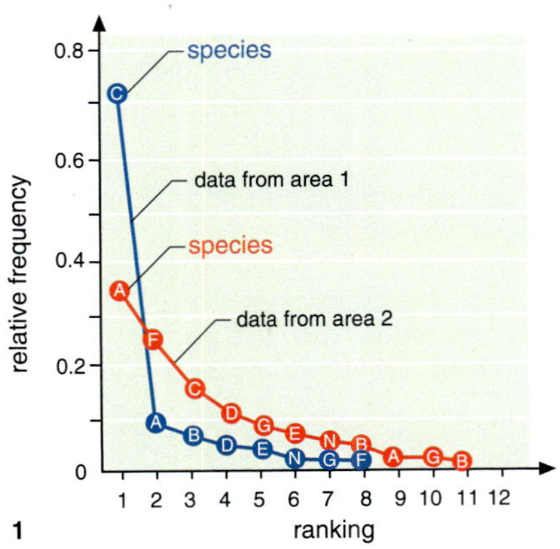

1

#### Tasks

⑦ Calculate the relative frequencies for the various species found and rank them in a table according to their frequency p. Add two empty columns (for task 10). Draw a graph of frequency (y axis) against ranking (see figures 1 and 2).
⑧ Compare the rank-frequency distributions for the different locations. What conclusions can be drawn?

**b) Diversity index according to Shannon**

The diversity of a biocenosis or its constituent parts can quantitatively be determined by the diversity index according to Shannon ($H_S$).
It is imperative:
$H_S = -\sum(p_i \cdot \log p_i)$
$H_S$ = diversity index
S = number of species
$p_i$ = relative frequency of species i

The equation takes the number of species and their frequency into account. Both rare and highly frequent species are weighted less than species of average frequency. The term p · log p allows this weighting.

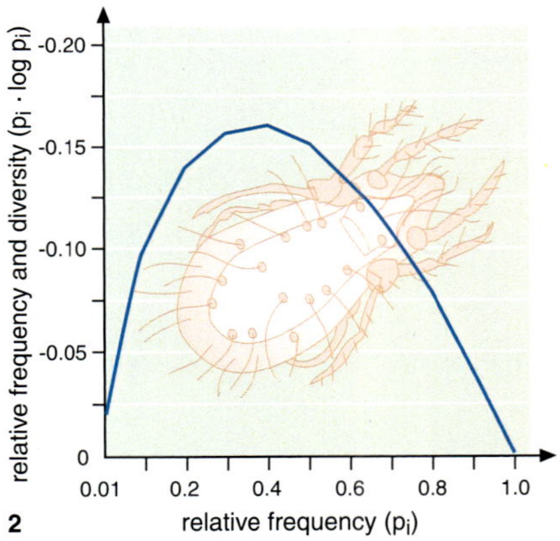

2

⑨ It is a general principle in ecology that, in highly structured environments with many ecological niches (see p. 15), biodiversity is especially high. Predict which of your areas of investigation will probably have the highest biodiversity.
⑩ To your table from task 7, add the values for log p and p · log p and sum these values: you will obtain a value for $H_S$. Compare these results with your prediction from task 9.
⑪ Is there a link between the values of $H_S$ and the rank-frequency distributions?

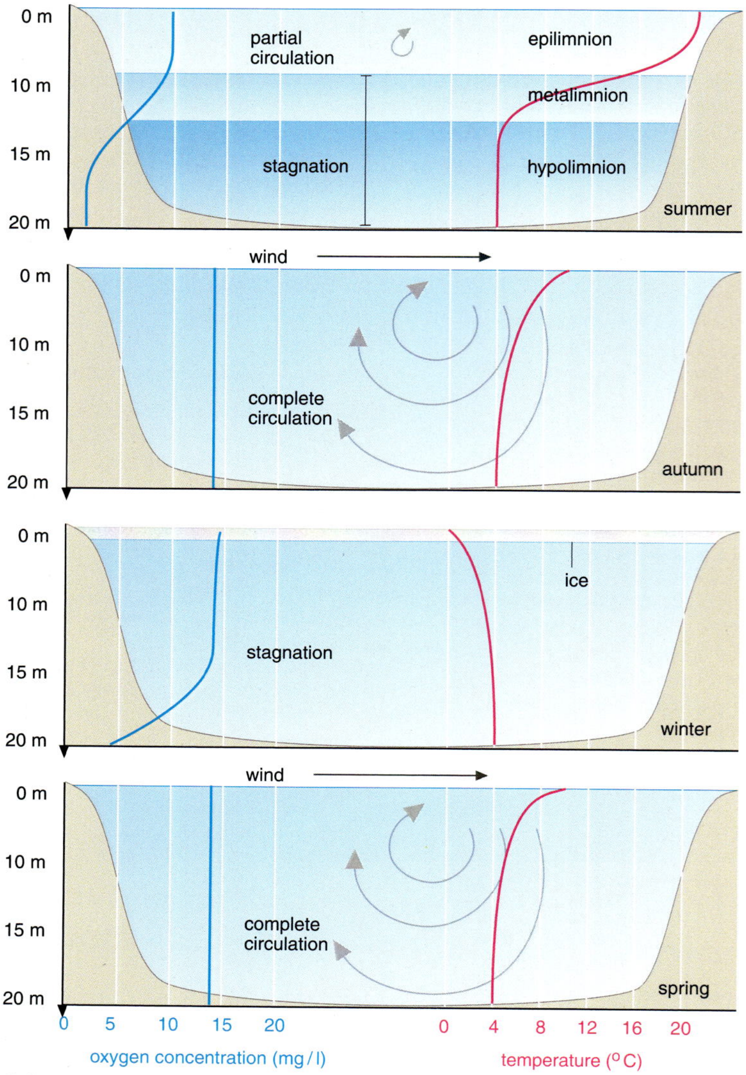


**1** Phases of stagnation and circulation during the year

## Lakes as ecosystems

The physical and chemical properties of water significantly impact on the possibilities of life in a lake. In addition to the oxygen concentration, which is strongly temperature-dependent, the density of the water is crucial (see info box). This causes the formation and disruption of *temperature strata* during the course of a year. In spring and autumn, the water in central Europe is of an almost uniform temperature; winds can produce a complete circulation of the water (*spring and autumn circulation*, fig. 1). Through circulation, the entire body of water is mixed, resulting in constant oxygen and mineral concentrations at all water depths. During the summer, a layer of warm water of a lower density (*epilimnion*) forms near the surface and results in a stratification of the lake. Solid particles can sink through the epilimnion into the colder and more dense layers of water below (*hypolimnion*), but no mixing of these strata occurs (*summer stagnation*), which means that the oxygen distribution is uneven; the water at the bottom could be considered as having had a lid put on it. In winter, a phase of stagnation can be caused by the formation of ice (*winter stagnation*).

During mild winters in general and for lakes in southern Europe in particular, no winter stagnation occurs and the water circulates from autumn until spring. The water of cold lakes, e.g. those in the Alps, remains cold even during the summer; circulation therefore continues also during the summer months.

## »info box«

### Physical and chemical properties of water

Contrary to most other substances, water has its highest density not at its melting point (0 °C) but at +4 °C. This *density anomaly* means that the water temperature at the bottom of lakes is never lower than +4 °C, as colder water always rises to the surface.

Ice also has a lower density than liquid water and therefore floats; it does not sink to the bottom like the solids of other substances.

Water is a *solvent* for many gases, e.g. oxygen and carbon dioxide. To some extent, water reacts with carbon dioxide to produce carbonic acid (dihydrogen carbonate), i.e. it can absorb substantial amounts of carbon dioxide.

The solubility of gases in water generally decreases with rising temperature.

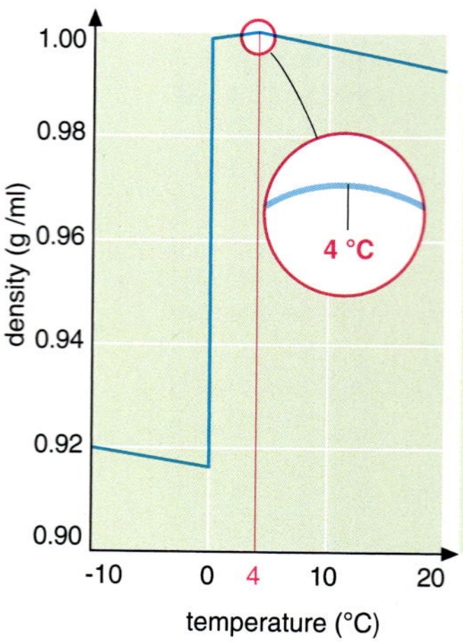


| °C | mg $O_2$/l |
|---|---|
| 0 | 14.2 |
| 5 | 12.4 |
| 10 | 10.9 |
| 15 | 9.8 |
| 20 | 8.8 |
| 25 | 8.1 |
| 30 | 7.5 |
| 35 | 7.0 |
| 40 | 6.6 |

oxygen saturation

As *incident* sunlight is absorbed by water, light intensity decreases with increasing depth. Up to a certain depth, production by plants through photosynthesis is higher than their respiration; beyond this depth, net production is impossible. This *compensation depth* (fig. 1) separates the upper and lighter layers of water (*photic zone*) from the lower *aphotic zone,* in which only processes of consumption can take place. The actual location of the compensation depth depends on the current light conditions and the amount of plants present. The layers of production and consumption depend on light, whereas the stratification of the water is, on the contrary, temperature-dependent.

The food web of a lake consists of the trophic levels of plant algae (phytoplankton). Above this, there is a layer of filter-feeding zooplankton. The third level contains fish feeding on the zooplankton (see centre column). If the productivity of a lake is sufficiently high, there can be a further trophic level of predatory fish. The bottom (*benthic zone*) is inhabited by snails, mussels and worms; larvae or imagi (adults) of many insects also inhabit water.

During the yearly cycle of a lake, life starts to proliferate in spring. With increasing light intensity and temperature, the phytoplankton populations start to grow. Following this, the populations of zooplankton also increase and consume, as effective predators (see p. 25), the phytoplankton until late spring. This can lead to a *clear-water phase*.

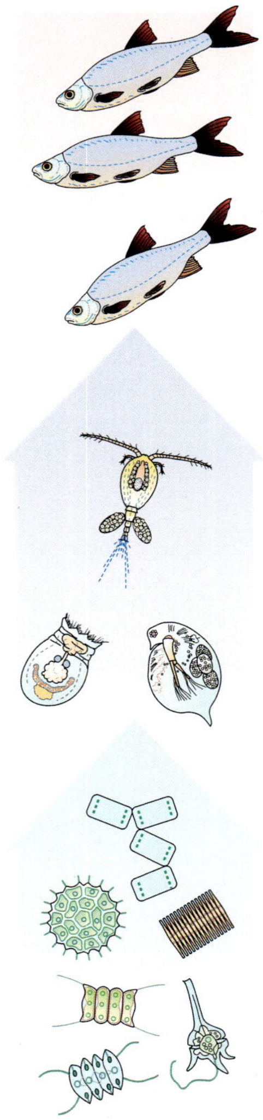

typical food chain

The further development depends on the remaining mineral concentration in the epilimnion.

Bacterial decomposers remineralise detritus and corpses using up oxygen. Most of these decomposition processes occur in the upper, oxygenated layers of water. Yet a considerable part of the dead organic matter sinks to deeper layers and, with it, the *mineral nutrients* that it contains; the upper layers of water slowly become depleted of minerals. If sufficiently high mineral concentrations remain in the upper layers, however, another proliferation of algae can occur (*summer proliferation*) (fig. 2); otherwise this is only possible in autumn when the mineral concentration increases again through renewed circulation.

Dead organic matter that has sunk to the bottom is also remineralised and hence, the mineral concentration at the bottom increases. This process uses up oxygen; in higher layers of water, oxygen can be replenished through diffusion from the air and also through photosynthesis. The stratification of the water during the summer prevents this in lower water layers or at the bottom. Here, oxygen is continuously used up and its concentration therefore decreases. It can run out completely if there are large amounts of *detritus* and therefore extensive decomposition processes (see p. 52). It is only with the beginning of the next complete circulation that new oxygen reaches the lower water levels and that minerals are brought back into the upper layers.

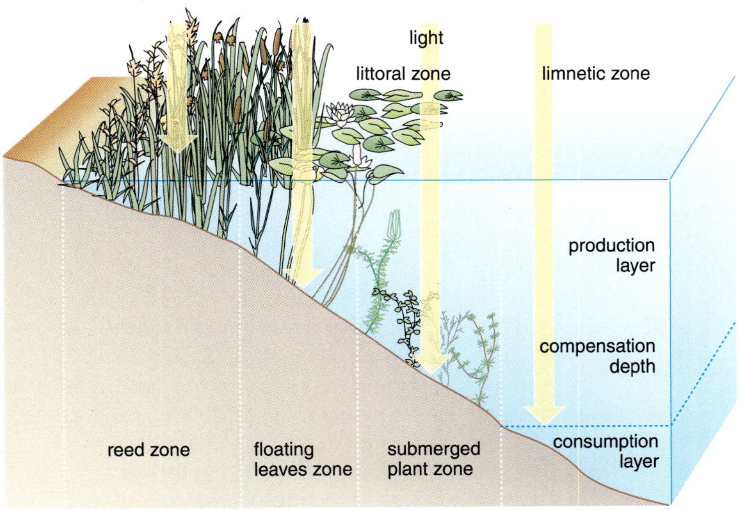


**1** Light-dependent layers in a lake

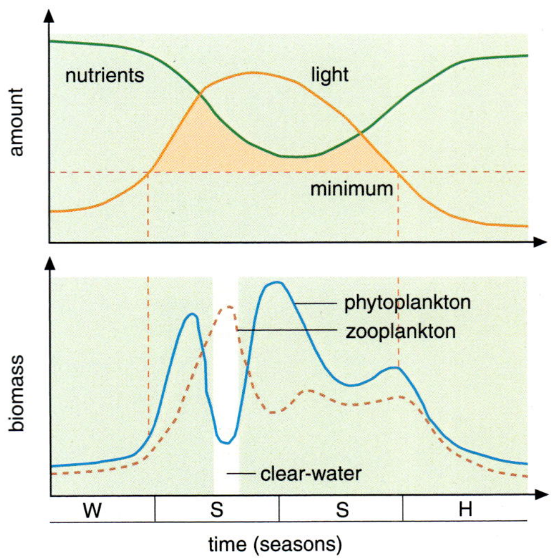


**2** Yearly cycle in a mineral-rich lake

1 Eutrophication in a lake

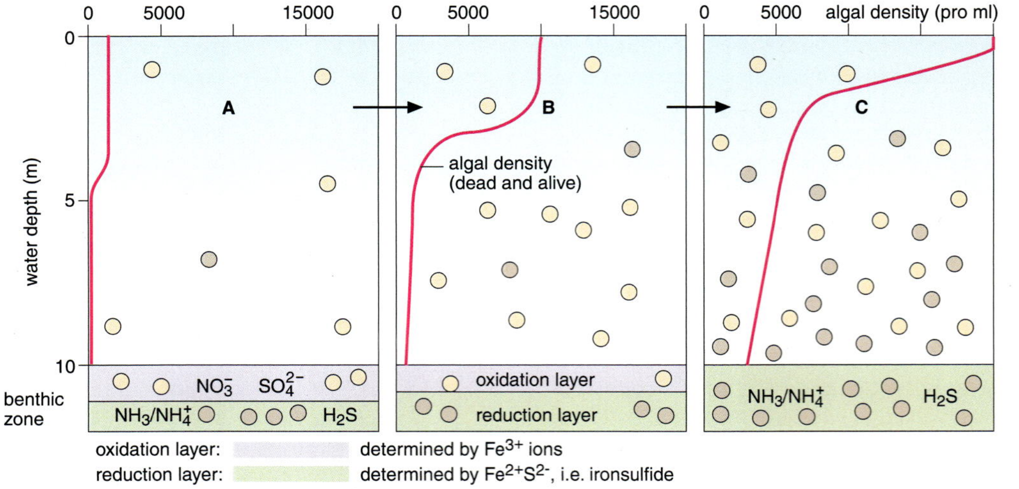


2 Processes occurring in a "collapsing" lake

## Eutrophication

Without human influence most bodies of water of considerable depth would be *oligotrophic* (mineral nutrient-poor): the concentrations of free *phosphates* and *nitrates* are low as most of these ions are bound in the biomass of living beings.

Through human (anthropogenic) influence, vast amounts of nitrogen compounds and phosphates can pass into open water. Agriculture is the main source as manure and *fertilisers* can be washed into lakes and rivers by the surface water. The effect of this mineral supply is *eutrophication*. The availability of previously limiting factors such as nitrate or phosphate increases the productivity and biomass of a lake, as is most evident by strong algae proliferation during summer ("algal bloom").

**Eutrophication**
Accumulation of minerals in a body of water

The *oxygen balance* of a lake is strongly affected by this eutrophication process: the decomposition of large amounts of biomass through microorganisms requires large amounts of oxygen. When there is a lack of oxygen, some organisms obtain oxygen through the *reduction* of nitrates and sulphates and so continue to oxidise organic matter to produce ATP (anaerobic respiration). These reduction processes produce substances such as ammonia ($NH_3$) and hydrogen sulphide ($H_2S$) (fig.3), which are highly toxic for vertebrates.

This results in the deaths of organisms that are oversensitive to these substances. More decomposable biomass is produced. If excess amounts of minerals are supplied, these processes can build up extremely quickly and can, within hours, lead to the deaths of all higher organisms; this is generally known as the "*collapse*" of a lake.

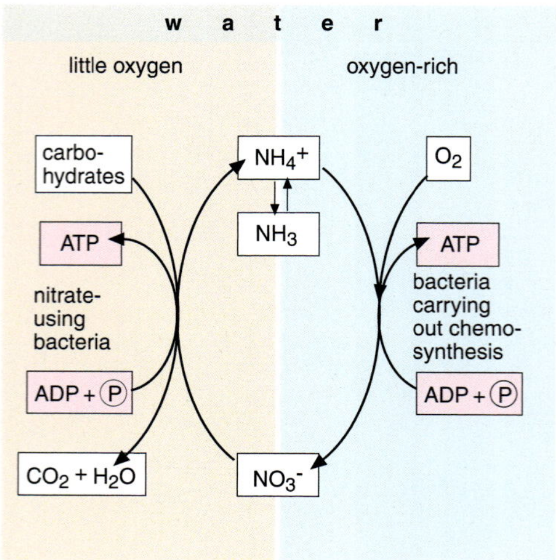


3 Reactions with nitrogen in microorganisms

### Tasks

① Try to find an explanation for the "collapse" of a lake. Use figures 2 and 3.
② Suggest reasons why the "collapse" of a lake occurs most often towards the end of a night during the summer months.
③ Suggest short and long term measures for the avoidance of the "collapse" of lakes.

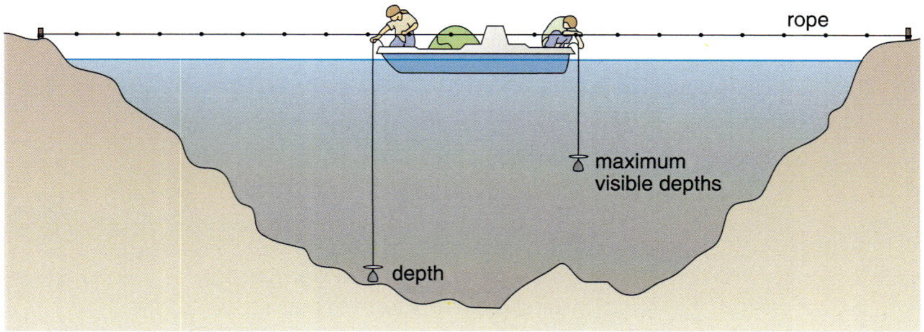


Measurement of the width and depth of a lake

## Outdoor ecology in a body of standing water

Ideally, choose a body of water that allows access to deeper water zones, e.g. via a gangway (landing stage) or boat.

### Vegetation record along a transect

Use a map to draw a sketch of the body of water and mark important features (e.g. trees) in your sketch for better orientation.

Draw up a profile of the depths along a chosen line. Start a few metres away from the water edge on dry ground.

Record the *vegetation* along this chosen line. Is there a zoning pattern according to height/depth?

### Physical and chemical data

Determine the water transparency along this line.

Take water samples from different depths by using the sampling bottle and analyse the water with regard to:

- temperature
- oxygen concentration
- pH
- concentrations of ammonium ions, nitrites and nitrates
- phosphate-ion concentrations

### $BOD_2$ analysis

The *$BOD_2$ value* represents the biochemical oxygen demand of organisms living in darkness. The oxygen demand over two days is used as the reference value.

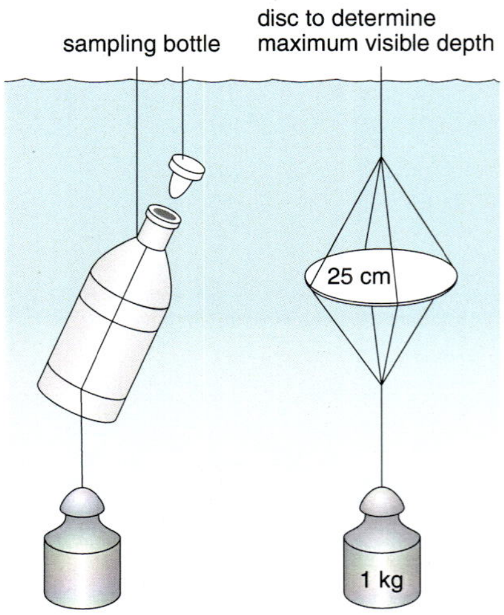


Completely fill a water bottle with a water sample. Measure the current oxygen concentration and seal the bottle. Avoid air bubbles. Place the bottles with your samples in a dark location at room temperature.

Boil the water of a different sample for about 10 minutes; once it has cooled down to room temperature, use the air pump of an aquarium to replenish the water with oxygen, then measure the new oxygen concentration. Now proceed as with the other samples.

Reopen the bottles after two days and immediately measure the oxygen concentration.

How has the oxygen concentration changed in the samples?

Compare with the changes of oxygen concentration in the boiled sample.

Establish a hypothesis to explain the net differences in the various samples.

Plan a lab experiment to test your hypothesis.

### Investigations over a day

Choose locations especially rich or poor in vegetation (water depths of about 20 cm) and measure the pH value every three to four hours. Plot the values for the locations on a graph and compare them. How can the changes during the day and at different locations be explained?

### Periphyton

In order to investigate the *periphyton* (small organisms attached to the submerged bottom), hang two opposing microscope slides at different depths into the water and leave them there for at least two weeks (see diagram). Try to choose locations in water with different light conditions. Ideally, place two or more probes at each location in the water so that you can continue your investigation over two-week intervals.

Examine the organisms on the slides under a microscope. Compare how the slides have been populated depending on water depths and, if possible, the light exposure at the location.

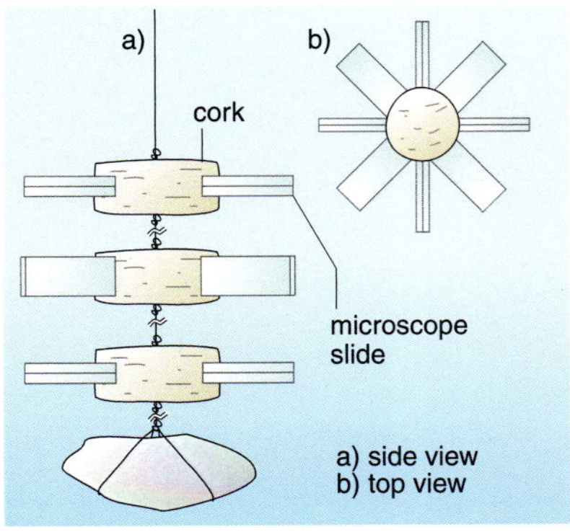

# Streams and rivers

Streams and rivers are characterised by the swirls in the water produced by the current. They oppose a stratification of the water and ensure a rich supply of oxygen.

Headwater streams often go through mountainous areas with steep slopes. The speed of a river therefore usually decreases from the source to the mouth. As a consequence, organisms such as water fleas and algae, which can also live in lakes, can be found further downstream. The headwater streams, on the contrary, are predominantly populated by organisms adapted to stronger currents.

**3** River in its natural state

| **Quality category** | **Degree of pollution with organic material (saproby index)** | **Chemical parameters** BOD$_2$ (mg/ml) | $NH_4^+$ (mg/ml) | $O_2$ minima (mg/ml) | **Important indicator organisms** |
|---|---|---|---|---|---|
| I | not / hardly polluted (1.0-1.5) | 1 | only traces | > 8 | freshwater pearl mussel, larvae of stoneflies 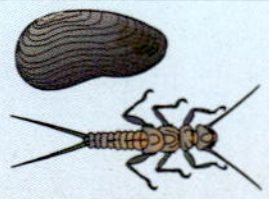 |
| I - II | minimal pollution (1.5-1.8) | 1-2 | approx. 0.1 | > 8 | larvae of caddisflies and stoneflies, freshwater shrimp, *Ancylus fluviatilis*, larvae of mayflies 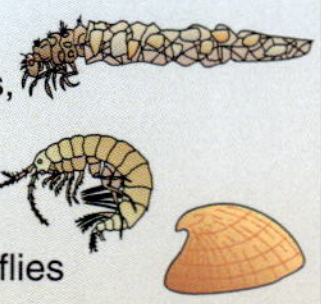 |
| II | moderate pollution (1.8-2.3) | 2-6 | < 0.3 | > 6 | |
| II - III | critical pollution (2.3-2.7) | 5-10 | < 1 | > 4 | leeches, snails, waterlouse, whirligig beetles 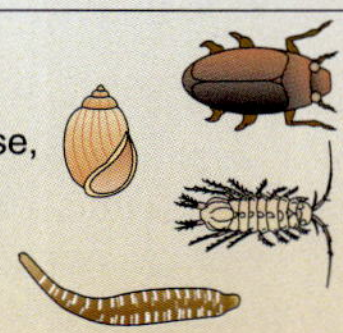 |
| III | strongly polluted (2.7-3.2) | 7-13 | 0.5 mg/ml and above | > 2 | |
| III - IV | very highly polluted (3.2-3.5) | 10-20 | several mg/ml | < 2 | Chironomidae, sludge worms, *Paramecium* 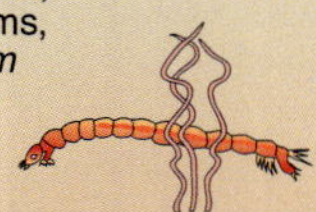 |
| IV | extremely polluted (3.5-4) | > 20 | several mg/ml | < 2 | |

**1** Saprobic system

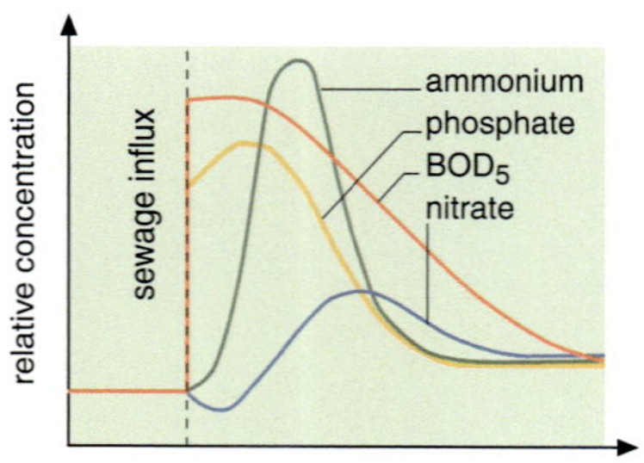


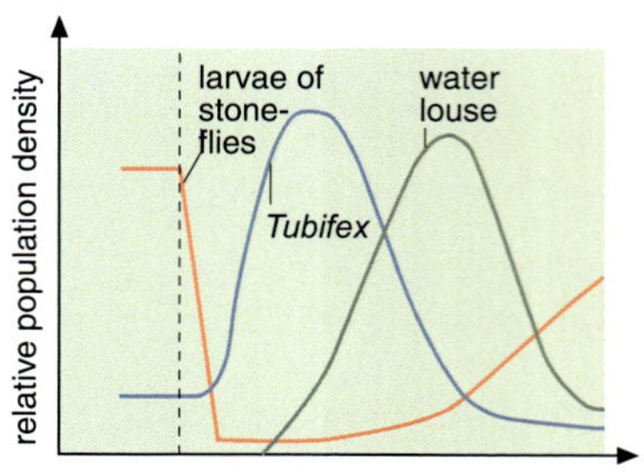


**2** Self-purification of a river / stream

The structure of a river is of decisive importance for its biodiversity: river sections that have been straightened and have a uniform current speed are usually poor in species, whereas meandering waters with their typical changes of undercut banks and sloping banks in close proximity can have markedly different current conditions and also show a variety of sediments on the river bed. Such structural variety promotes the biodiversity of streams left in their natural state (see p. 45).

## Self-purification and saprobic systems

The biggest threat for rivers and streams lies in the addition of organic substances such as liquid *manure* from agriculture. Decomposition of these substances through microorganisms uses a large amount of oxygen and similar to lakes, in adverse conditions this can lead to the "collapse" of the river (see p. 52). Other microorganisms later oxidise these substances and form less toxic salts such as nitrates or sulfates, again using a large amount of oxygen. This *self-purification* of rivers and streams can be enhanced by a better supply with oxygen, which can be achieved through e. g. weirs. The discharge of warm water into rivers by power stations or *pollution with heavy metal ions* (industrial sewage) are problematic factors for the self-purification process.

As water keeps flowing during the process of self-purification, the progress of decomposition over time can be seen along the *spatial zonation* of a river (see fig. 2).

**Eutrophication**
Accumulation of minerals

**sapros**
(Gr.) = rotten, decayed

**bios**
(Gr.) = life

**Saprobes**
Organisms inhabiting polluted water

Inorganic salts such as phosphates and nitrates that originate in agriculture (fertiliser), industry and private households, contribute significantly to pollution (*eutrophication*, see p. 52). Because of an ample supply of food, a large number of individuals usually live in polluted waters, although the number of different species is low as only a few organisms can tolerate a low concentration of oxygen and high levels of ammonium.

Regular checks of the water quality are extremely important; the $BOD_5$ value is a crucial indicator. It measures how much oxygen of a water sample previously saturated in oxygen is used up through respiration by microorganisms over five days, at a temperature of 20 °C and in complete darkness (cf. $BOD_2$, p. 53).

A biological way of judging water quality are *saprobic systems*, a method that has been used for more than 100 years. The degree of pollution is determined by the presence/absence of chosen *indicator organisms*. The degree of pollution determines the quality category of any section of a river or stream.

**Tasks**

① Consider the following parameters: $BOD_5$, ammonium/concentration of ammonia and oxygen (fig. 2). Explain how these parameters are linked.

② What consequences would you expect if the temperature of a river was increased, e.g. through the discharge of cooling water from a power station.

③ Compare natural self-purification process with purification in sewage plants.

»info box«

## Working of a sewage treatment plant

The treatment of domestic sewage from the public sewage system (1) happens in several stages. The *first mechanical treatment* (primary treatment) removes large objects from the waste water by rakes and sieves. Small particles suspended in the water are disposed of as sludge by decomposition in especially designed *digestion chambers* (9). During the second phase (*biological treatment*), organic substances in the waste water are remineralised; this reduces the concentrations of nitrogen compounds and phosphates. The shortage of oxygen in the *denitrification tank* enables specific bacteria to reduce nitrates into nitrogen $N_2$. The remaining nitrogen compounds are oxidised by bacteria into nitrates. This happens in tanks where oxygen is supplied in excess (activated sludge) (6). The nitrates produced here can then be returned into the denitrification tanks so that, in the end, most of the nitrogen compounds are changed into and released as gaseous $N_2$.Phosphate compounds are transformed into soluble forms in the denitrification tanks. In this form they can be ingested and fixed by bacteria. These microorganisms and other protozoa, which also feed on the organic material, *flocculate* and can be collected as sediment in the final sedimentation chamber (7). Some of the flocculated material is used to reinfuse the activated sludge with microorganisms. Most of the flocculated material (sludge) is decomposed in *digestion chambers* where bacteria decompose it under anaerobic conditions, thereby producing *methane*. The remnants after the decomposition process can be used as fertiliser or are disposed of in landfill sites if contamination levels are too high. The purified water is discharged into a *runoff ditch* (8).

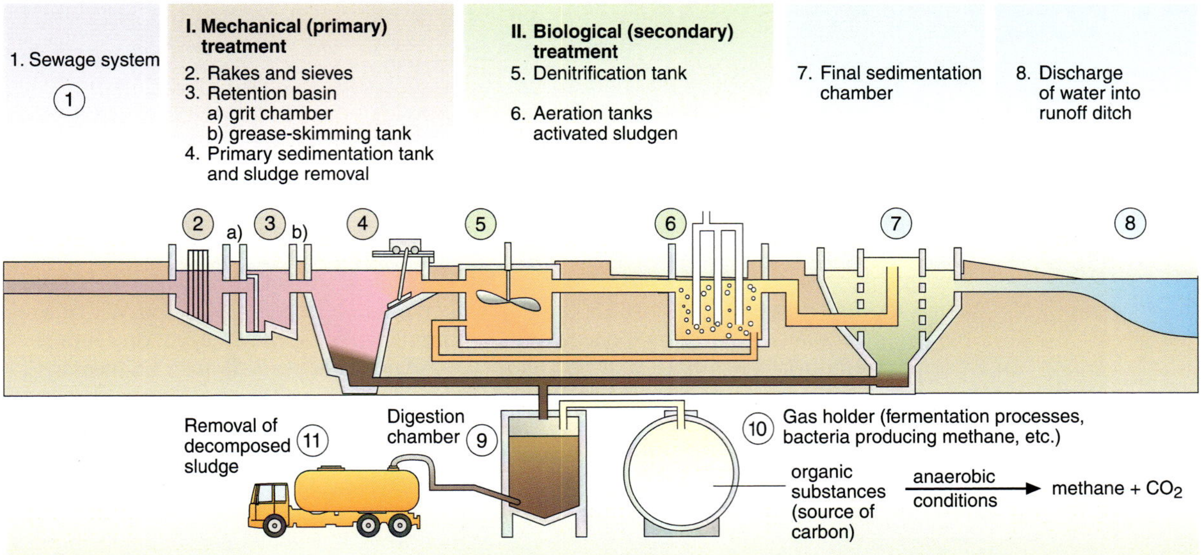

## World population

As seen from space, Earth is obviously a living space of limited dimensions. The number of people living on this planet however is steadily rising (fig.1). Every day about 250 000 additional people need space to exist, homes in which to live and agricultural land to ensure their food supply. Given these growth rates, it will not be possible to feed all these people for much longer; in the long run many ecosystems will be destroyed, which will mean a reduced quality of life for large parts of the world population. Apart from food shortages for many people, health problems will arise because of air and water pollution.

With the introduction of agriculture and farming, humans renounced the position of hunter-gatherers and the world population started to increase. Further developments in agriculture and farming, the use of fertilisers and effective *irrigation* methods were a prerequisite for a thousand-fold increase in the overall yields. The landscape started to change now that it was economically used for agriculture and provided entirely new living conditions — one of the prerequisites allowing for the large world population that we have today. Humans started to create areas in which only consumption of goods took place — towns; they became separated

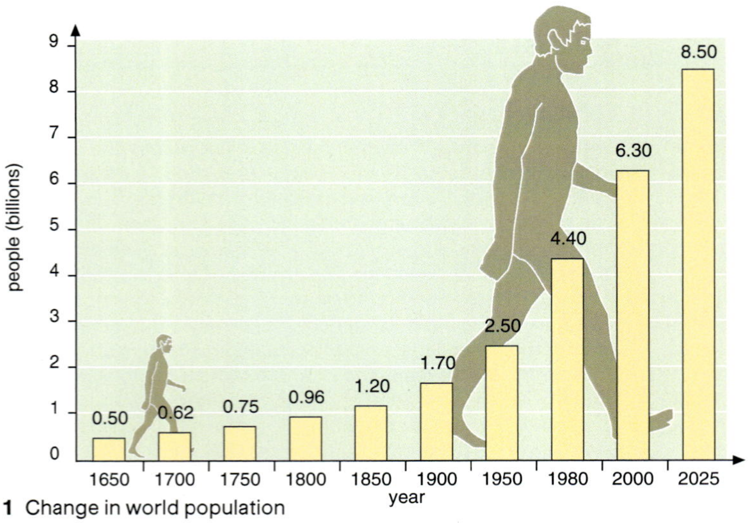

**1** Change in world population

**2** A problem or just a cartoon?

The history of how humans populated the world shows that the unlimited growth of population density is impossible. When humans were *foraging* the forests as hunter-gatherers, they were part of their ecosystem as consumers. A maximum of 3 to 5 people could share one square kilometre and find enough food. Under these circumstances, a maximum of roughly 30 000 to 50 000 people could have lived on an area the size of Germany. Historians estimate that the entire world population at that time was between 1 and 10 million people in total. There simply would not have been enough resources (food and places for settlements) for additional people.

from the areas of production, i.e. the fields and forests. This separation led to, by now, worldwide trade and an ever increasing energy demand for the transport of the goods produced. The limits of population density are set by consumption, environment-friendly technology and energy-use. The overall area of ecologically used land is estimated at about 9 billion ha. With a world population of about 6 billion, on average every individual would have 1.5 ha at their own disposal.

The distribution is however extremely uneven; for the living standard in India, 0.4 ha are required per person. Europeans would need 3 to 4 ha and US Americans would need 5.1 ha each.

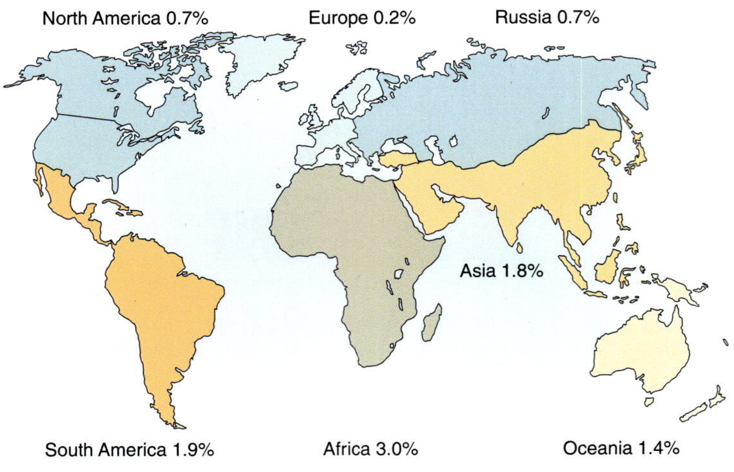


**1** Populaton growth rates

Assuming an environmentally sustainable and modest life style, which requires 2 to 3 ha per person, a maximum of 3 to 4.5 billion people could fit onto this planet.

The growth rates of populations in the various regions of the world in the years between 1990 and 1995 show major differences (fig. 1). In Europe, the population is only growing by 0.2 %, whereas in Asia it is going up by 1.8 %. Using the actual numbers of citizens in these regions as a basis, this means that about 60 % of the overall additional growth is expected to occur in Asia.

The different developmental pattern in these regions has a number of reasons. Development among industrialised nations is determined by relatively high income levels, improved social services, a high level of education, the availability of *contraceptives* and good and effective medical provision. Mortality has decreased, especially through advances in medicine. Antibiotics and other medicines have minimised or even eradicated fatal infectious diseases and epidemics. A high income and social services such as pension schemes help to ensure a good level of general provision in old-age.

Moreover, birth rates have dropped as people are no longer dependent on the care of their children. Technological advances in domestic appliances and an improved general and vocational education have enabled women to have jobs and to be more economically independent. In connection with birth control, this has led to lower birth rates.

Among developing countries, however, the situation is different. The majority of people are poor and there are no institutionalised provisions for the elderly. Another factor is early marriage, sometimes even during the teenage years, which significantly prolongs the time during which women can become pregnant. The success of educational advertising campaigns for family planning and birth control therefore remains limited in parts of the world where child mortality is high and where few or no social provisions are available. On average the population is very young, with about one third being 15 years old or younger. Even if birth rates can effectively be curbed in the coming years, the world population will continue to increase for another few decades to come, as many young people move into adulthood and become sexually mature.

Evidence suggests that, with a higher level of general and vocational education in developing countries, women become more interested in deliberate family planning. Parallel to information campaigns about birth control, ways need to be found that secure provisions for the elderly who then become independent of their children for support; only in this way can a reduction of population growth be successfully achieved.

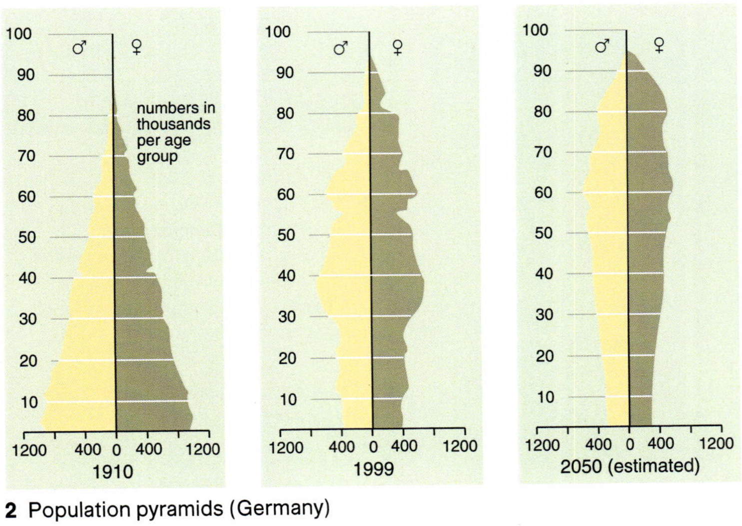


**2** Population pyramids (Germany)

## Shield around the earth

Our planet has a protective shield around it — the *atmosphere*. It surrounds the world like a *veil* and is about 100 km thick. Clouds are found in the layer nearest to the ground, i. e. in the 10-km thick troposphere. This layer is responsible for the climate on the Earth's surface. It is of importance to animals and plants because it provides protection from cosmic radiation. Ozone is especially important here as it absorbs UV radiation. Furthermore, the atmosphere helps to retain some of the heat radiation that is reflected from the Earth's surface. The release of different gases such as carbon dioxide changes the composition of the atmosphere. Other gases released from technological processes sometimes also play a role.

### Ozone in the stratosphere

„Between eleven and three, slip under a tree".

The ozone layer in the stratosphere acts like a filter and protects the Earth's surface from short-wave electromagnetic radiation emitted by the sun, e. g. UV light. Increased UV radiation can damage the DNA and proteins of all living organisms.

The area of strongest ozone *depletion* is found above Antarctica. This hole in the ozon layer extends as far as Australia. Among citizens of Australia and New Zealand, there is a correlation between the occurrence of skin cancer and ozone depletion.

**Provide a summary of the chemical substances involved in the depletion of ozone in the stratosphere. Find information concerning the origin of these substances, i. e. whether they are produced in either natural or technological processes and explain their significance for the process of ozone depletion.**

**Do we face the same danger in Europe? Is there an ozone hole above the Arctic?**

**Find information and data regarding these questions and write a newspaper article commenting on this data.**

**Up-to-date data about ozone can be obtained from the ozone monitoring experiment (GOME) at the University of Bremen: www.iup.physik.uni-bremen.de**

### Ozone at ground level

On a beautiful summer's day, often no blue sky can be seen above a city. A layer of nitrous oxides and ozone covers the city in smog.

The release of pollutants into the atmosphere is called *emission*. Emissions that have an impact on living organisms or buildings are called *immissions*.

For ozone, the *maximum immission concentration* (MIC) at ground level is 120 micrograms per $m^3$ of air.

Below this threshold level, even ozone-sensitive people should not feel any negative impacts on their well-being, despite being exposed for prolonged periods of time. Smog warnings are given once ozone exceeds a concentration of 240 micrograms per $m^3$ of air.

**What specific health problems can be caused by ozone?**

**To what extent have improvements in car technology helped to reduce the danger of smog?**

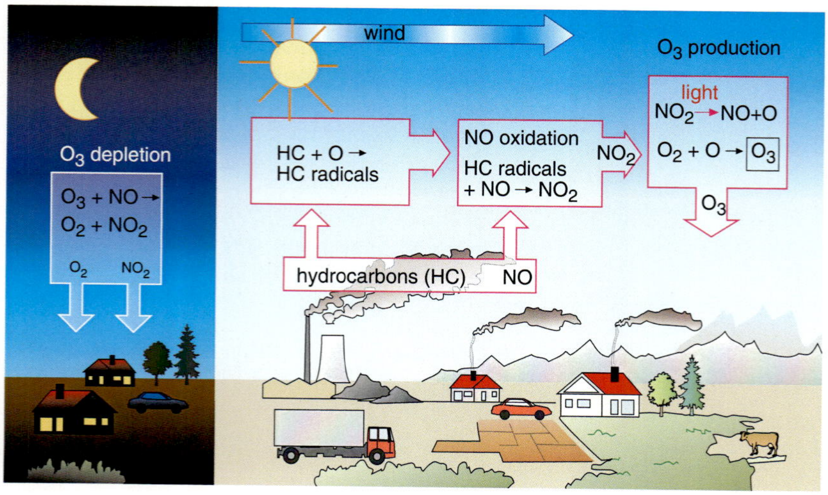

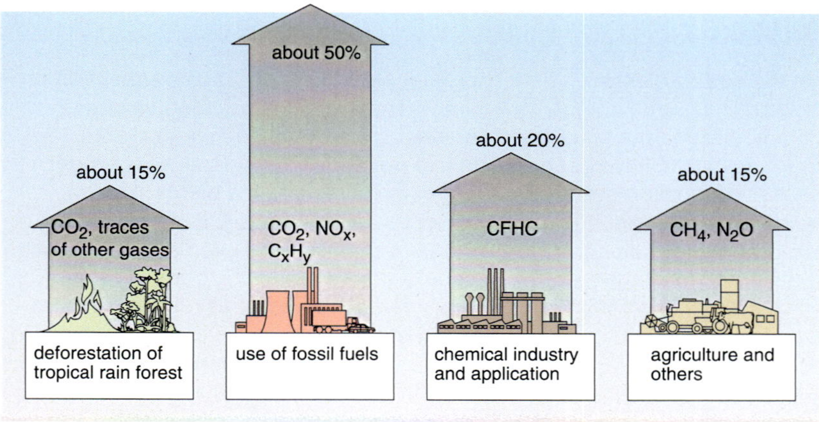


## The greenhouse effect

Everyone knows that the inside of a car heats up rapidly if the sun is shining in through the windscreen. This effect is also used in greenhouses, hence the name *greenhouse effect*. More energy shines in than is emitted. The temperature remains constant only once the incident energy is equal to the energy emitted.

| Global annual emissions of methane in millions of tons per year | |
|---|---|
| Swamps | 121 – 205 |
| Rice fields | 70 – 170 |
| Animal farming | 80 – 100 |
| Termites | 25 – 150 |
| Industry | 5 – 25 |

The Earth's atmosphere corresponds to the glass of a greenhouse. Water vapour, carbon dioxide, and methane absorb heat. This natural greenhouse effect maintains the average temperature of the atmosphere at around +15 °C. Without this effect, the temperature would plummet to -18 °C.

Human activity is changing the composition of the Earth's atmosphere. The concentration of carbon dioxide has increased over the last 200 years by about 25 %. The temperature increase is also helping to thaw the permafrost soils and frozen swamps (tundra) in Siberia and methane gas, which is the trapped there, is beginning to escape.

Predictions of the precise effects that this will have on the world's climate are impossible to make. On the basis of model calculations and simulations, researchers try to give an idea of possible future developments: more storms, increased sea levels, alterations of desert locations and changes of climatic zones are being forecast.

**The greenhouse effect is judged differently. Collect information regarding the different viewpoints and compare them.**

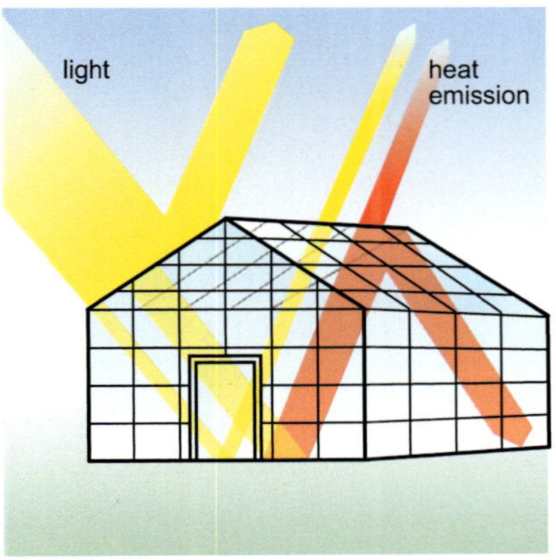


## Renewable raw materials

Bio-diesel is one way of directly using a raw material produced by plants, i.e. rape seed oil as a fuel for combustion in diesel engines. For this purpose, the rape seed oil is separated into *glycerol* and fatty acids. The latter are reacted with methanol to produce methyl esters; this material has similar properties to those of diesel and can therefore be used as a fuel-substitute in diesel engines.

Bio-diesel contains about three times the energy invested in its production. In order to reduce $CO_2$ emissions in Germany by 1.4 %, 10 000 km² would need to be used for growing rape seed.

The glycerol obtained in the above-mentioned refinery process is an important raw material for the chemical industry, e. g. for the production of cosmetics.

After the oil has been pressed out from the seeds, the remains (so-called rapeseed cake) contain cellulose, about 9 % fats and 34 % proteins and are high-quality animal feed.

**Find additional information regarding bio-diesel and write a report for or against its use.**

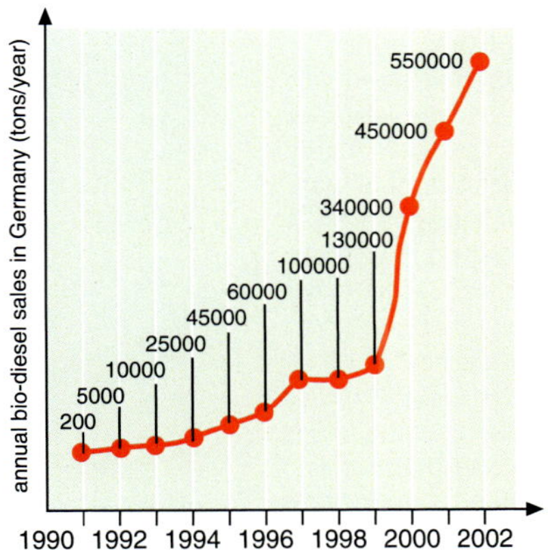

# Endangered forests

History shows that the destruction of forests through humans has been going on since ancient times. Around the Mediterranean, whole stretches of land have been cultivated and devastated through overexploitation. In areas with heavy industry, the damage to trees caused by sulphur dioxide has been observed for more than 100 years.

During the 1970s, some fir trees in Germany's lower mountain ranges were first observed as starting to show yellow needles and bare twigs. Later such damage was also observed on spruces and pines growing in valleys. Since 1980 beech trees and other deciduous trees have also been damaged; the symptoms are always the same and are summarised as the "*forest dieback*". First the leaves change colour and then older branches start losing their leaves. The treetops become less dense and finally the whole tree dies (see p. 61).

This more recent damage to forests has been scientifically observed since 1982; the degree of damage is measured by leaf/needle loss compared with a healthy tree (see p. 61). Car exhaust fumes pollute the air with carbon monoxide, nitrous oxides and soot particles (*emission*). These pollutants are either gases or are deposited as solid particulates on the ground or on plants (*immission*). Some substances from the exhaust fumes can continue to react chemically and form *secondary pollutants* such as sulphuric or nitric acid, both of which are precipitated to the ground as either rain or snow. They either affect needles directly or indirectly through the plants' roots. The increased supply of minerals via nitrogen deposition changes the plants' growth, whereas ozone affects the opening of the stomata of leaves and needles. Damage also occurs in areas with little or no industry. Tall *smokestacks* of factories help to avoid pollution of nearby areas, but enable pollutants such as soot particulates, sulphur dioxide and nitrous oxides to be distributed further away (*transmission*).

Measurements in forest soils reveal that aluminium ions that are usually bonded to soil particles, are released in high concentrations through acid deposition. Scientists assume that this inhibits the absorption of magnesium ions. Magnesium is an important element for the formation of chlorophyll; insufficiency in the supply of magnesium directly results in a reduced rate of photosynthesis. Experimental data confirms a reduced absorption of magnesium ions, whereas other ions such as potassium are not affected. This deficiency, which can be considered as an additional stress factor to natural effects such as drought, frost or pests, weakens the trees.

The most efficient protective measure for stopping or at least slowing down the acidification of forest soils is to treat wide areas of forest with calcium magnesium carbonate; helicopters are often used to spread this chemical over large forest areas. In this

**Thermal (IR) image of a tree**
The thermal image of a tree taken at night shows that parts of the trunk and the branches are at different temperatures. The reason is the different amount of water that is transported through the vascular bundles. In healthy *vascular bundles*, large amounts of water are transported and the tissue is warmer than similar tissue transporting less water.

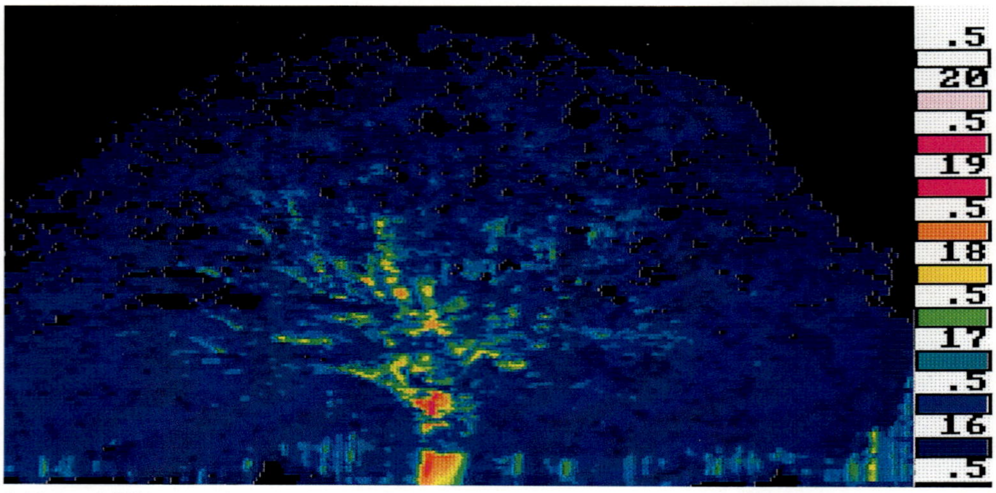


**1** Thermal (IR) image of a tree

K-deficiency (beech)

K-deficiency (oak)

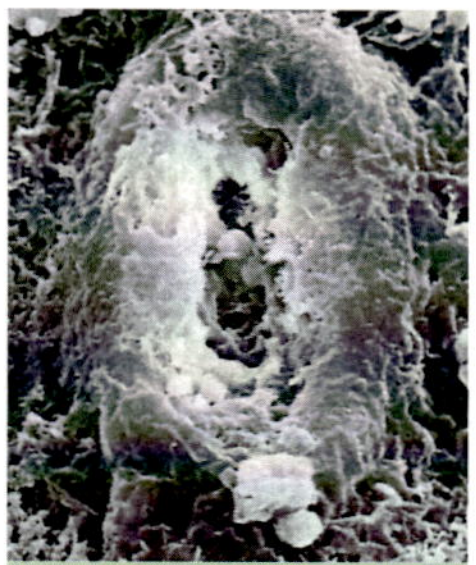
healthy cuticle

damaged cuticle

healthy roots

damaged roots

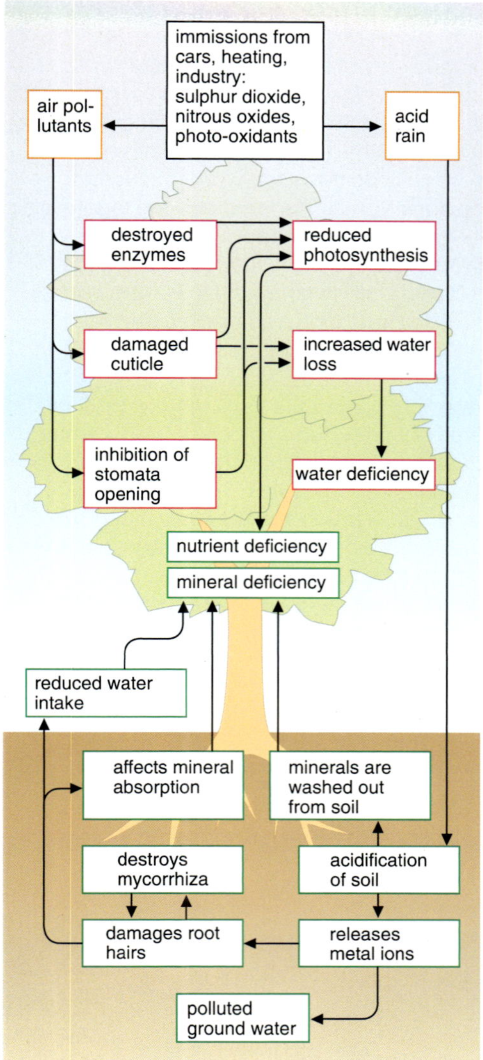

fir degree of damage: 0

degree of damage: 2

degree of damage: 3

manner the acids deposited in the soils are neutralised, the amount of magnesium needed for leaves and needles is increased and the basic minerals are replenished in the forest.

To a lesser extent, some non-domestic tree species are used for the reforestation of damaged forests, the douglas-fir from North America being one example. It is a more robust tree and is less sensitive towards damaging environmental influences. However, the introduction of non-domestic tree species should only be used if this does not have negative effects on the overall forest biocenosis and only if the new trees can successfully be integrated alongside with domestic tree species.

The stability of forest ecosystems in the long term can probably be achieved by aiming for naturally mixed forests and forest edges with a high biodiversity. It is only in these naturally evolved environments that the survival of species is ensured in the long run. This also helps to maintain the genetic variety and results in a stable ecosystem.

### Task

① Explain the effects of calcium magnesium carbonate on trees.

## Industrialised agriculture

corn poppy and cornflowers

Huge fields of corn as far as the eye can see. A *combine harvester* harvests the grain along long lanes. Every now and then the grain is loaded onto a trailer. Only one person is doing all the work that used to be done by many farmers working together. In 1930, about 149 hours of work were invested into each hectare of land; today it is less than two hours. Nowadays agricultural farms are on a similar technological production level as industry.

A variety of reasons have resulted in this type of agriculture. The cost of labour has risen much faster than the producer price. This difference could be compensated only by rationalisation and mechanisation, which inevitably led to structural change in many agricultural businesses. The number of farms has declined in recent years while the size of farms has more than doubled between the 1950s and today. Increased performance has only been possible through more machinery, which in turn is only economically viable on enlarged fields or animal stocks. The increased performance becomes evident in a comparison of farm productivity over the past decades. Whereas, on average, in 1950 a farmer produced food for about ten people, in 1990 the number had increased to 89. Over the same time period the annual food consumption increased from 800 kg to 1300 kg per person. The producers' prices dropped dramatically at the same time (see margin p. 63) and were only achievable through the intensive use of machinery.

The increase in meat consumption in the industrialised countries is an important ecological aspect (fig. 63. 2). This "improvement in human nutrition" means that more herbal feed needs to be grown than would be required if people's diet was predominantly vegetarian. This correlation can be explained by reference to trophic levels: only a small percentage of the sun's energy is transferred from the producers (i. e. plants) to the 1st order consumers (herbivores such as pigs or cows) and then to 2nd order consumers, i. e. humans (see p. 34).

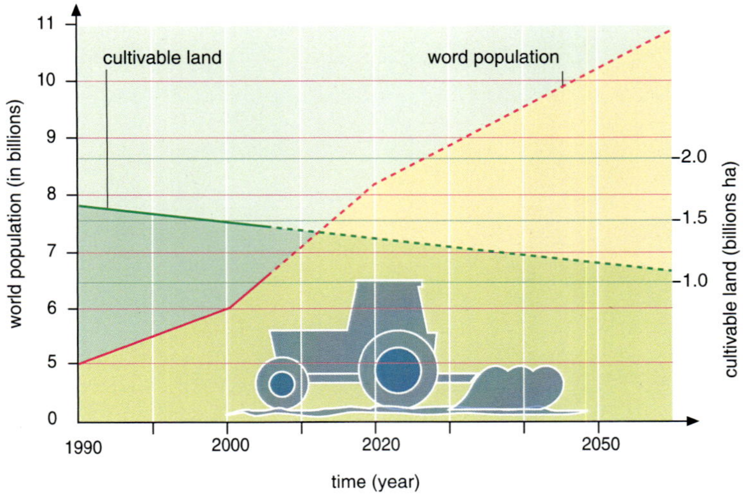


**1** Farm land and world population

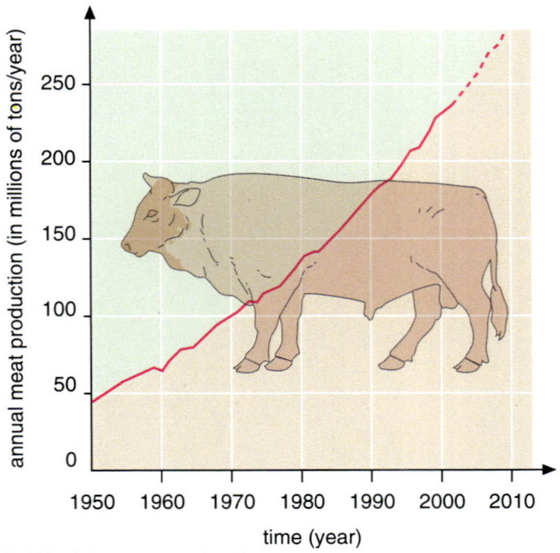


**2** World meat production

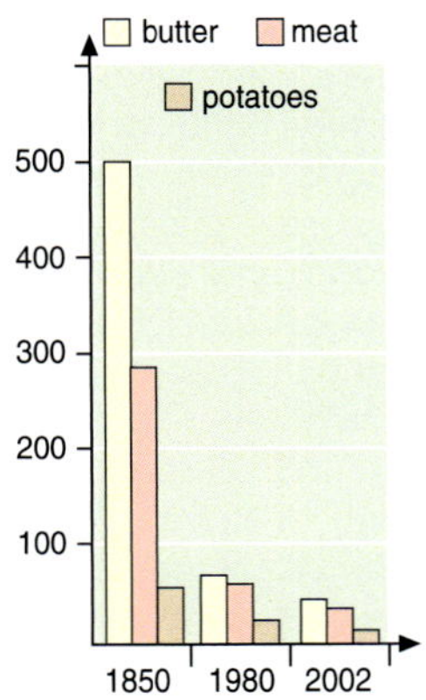


working time for 1 kg of food in minutes

*Plot clearance* occurred together with the formation of large-scale farms; whereas this allowed more economical production methods, it also led to an ecologically poor style of agriculture. Small parcels of land were joined to make large areas of cultivable land that were suitable for large farming machinery. Hedgerows, small trees or wetlands were lost. The number of animal and plant species living in the cleared areas decreased. Hedgerows, for example, are living spaces for birds, ladybirds and green lacewings, which helped to reduce the mass proliferation of pest insects.

Large numbers of one plant species in a coherent living space promote the proliferation of pest organisms. Pest control of insects, fungi or *weeds* nowadays is not performed mechanically or through crop rotation, but with highly effective poisons, which help to keep production costs down. *Pesticides* are highly effective substances that are spread onto fields. They need to be applied in the right quantities and *precautionary measures* must be taken (breathing protection, protective clothing) in order to avoid human health problems. The *accumulation* of toxic substances in the fatty tissues of plants or animals via the food chain is another hazard. Close observation of the degree of pest infestation allows the targeted application of specific pesticides. Plants that are genetically modified so that they are resistant to certain fungi or other pests are another new development. By focusing on only a few selected plant species, the genetic variety between plants is reduced and farmers can only use particular types of seeds and their follow-up products such as pesticides from a specific manufacturer.

Livestock farming in Europe is now impossible on open farmland because of the large numbers of animals and the high demand for meat. The demand for cheap meat can only be met by its production in barns with an exactly calculated food supply and a minimum of man labour. Many barns nowadays have heating facilities to promote an increased transfer of the energy taken in by the animals as food into meat and fat, rather than into body heat. Diseases spread easily in these "monocultures". This can be combatted through extremely hygienic conditions and the use of medication.

**»info box«**

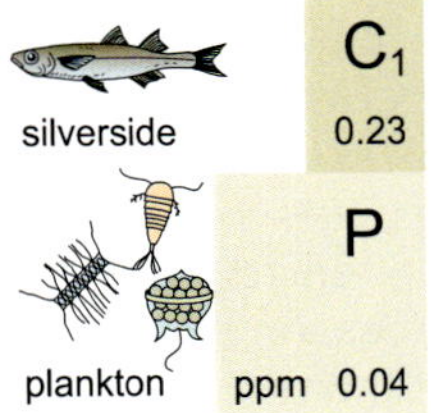


## DDT – (not) a miracle cure

The insecticide DDT was the first effective ingestible and contact poison produced on an industrial scale. Its advantage lies in its high compatibility with plants; its disadvantage however is its high molecular stability, which makes it accumulate in natural food chains. For vertebrates it only has a low toxicity but, as it is stable, it accumulates in their fatty tissues. Because of DDT, birds of prey had a lower reproduction rate, as DDT caused their egg shells to be thinner than usual. No immediate health effects have been established for humans. The application and production of DDT is now forbidden in many countries but, as DDT is cheap, it continues to be used in poorer countries to combat malaria (which is transmitted by the mosquito).

1 Moorland and its inhabitants

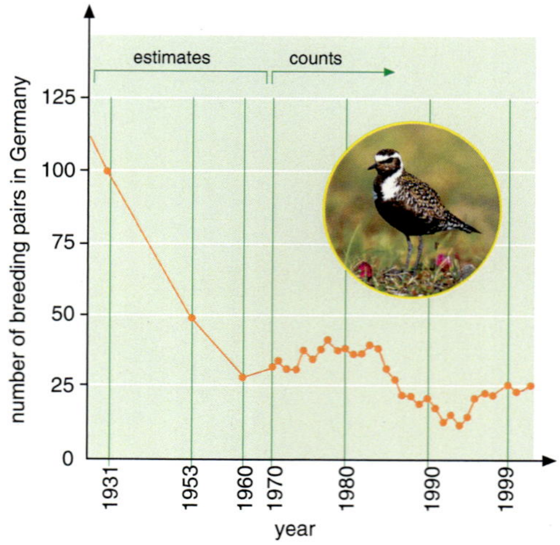


2 Population numbers of the Eurasian golden plover

## Nature conservation

common sundew

common cottongrass

common butterwort

Throughout the history of the world species have become extinct, but no species has ever caused and increased the speed of the process of extinction more than we humans today. Every day species die out in the tropical rain forests even before they have been discovered. In previous centuries, the main reason for *extinction* was usually hunting. Today the majority of plants and animals are endangered by the destruction of their environments. The protection of ecosystems and their biodiversity is at the heart of nature conservation. *Laws for nature conservation* ensure the protection of living spaces by imposing access *bans* restricting their use for agricultural and *recreational purposes*.

Such protection is often not enough, as the example of central European *moorlands* shows. These environments are characterised by the absence of trees, high humidity levels, a low pH value and a shortage of minerals. To make the surrounding areas economically usable, *drainage systems* are commonly put in place. As a result, birch and willow trees start to populate the moorland. Species such as the *black grouse* or the *Eurasian golden plover* now avoid these areas as they provide cover for predators. The surface areas of these pioneer trees lead to increased water loss through transpiration, which lowers the groundwater levels and allows even more trees to populate the area. As the emerging forest is an unsuitable habitat for typical moorland animals, they start to die out.

Habitats lacking minerals are especially endangered by *eutrophication* (influx of minerals) via the air or ground water: more competitive plants such as grasses or dandelions can displace weaker plants such as the *marsh gentian* or *sundew*, which are adapted to a low mineral supply. Insects such as the *alcon blue*, which depends on the marsh gentian as their food source, can no longer survive as a result. Walkers with no knowledge of biology usually remain unaware of these changes as most of these species are hidden and inconspicuous. Only the education of both the public and politicians about ecological interdependencies and the hidden biodiversity of endangered ecosystems will help to bring about an understanding for the problems of nature conservation.

Many of the habitats and living spaces that we want to protect today have only been created by the interference of humans in their efforts to cultivate an area. Originally from Asia, *skylarks* could only settle in Europe once large areas had been deforested for agriculture. Gravel pits and quarries are *secondary environments* and are now populated by a large number of endangered species. Many of these ecosystems are unstable and can only be preserved through intensive *maintenance*. *Grasslands with orchids* are one example, as these areas can only be protected from unwanted population by shrubs through non-intensive agricultural use. Nature conservation therefore also depends on cooperation with agriculture.

# Protection of wading birds

Wading birds are species that live in open landscapes. While the *common redshank* predominantly populates the salt marshes of the North Sea, which are otherwise only grazed on by sheep, all remaining species prefer wet and humid meadows inland. The populations of all wading birds mentioned here are highly endangered in central Europe. Their numbers have rapidly declined over the last 50 years, especially the *black-tailed godwit* and the *ruff*. Research into the disappearance of species shows strong links to agriculture.

## Requirements of wading birds

1. The vegetation should not be too high as, like most meadow birds, they need to be able to watch out for potential predators (fox, goshawk). This is especially true for the short-legged lapwing (also called the peewit or plover).
2. The vegetation must not be too dense, as otherwise young birds would be unable to move through the dense grass. This would make searching for food impossible.
3. The ground must not be too hard as otherwise their food sources (small invertebrates living underground such as the larvae of insects) would be inaccessible.

**1** Water meadow

### Tasks

① Intensive use of fertilisers, be it mineral fertiliser or liquid manure, results in the migration of wading birds. This also occurs with respect to loamy soils that are flooded during winter and early spring, but then fall dry in late spring because of artificial *drainage systems*. Explain this migration.

② Lapwings breed increasingly often in fields, but breeding success here is extremely low. Explain.

## Breeding success and cow density

The breeding success of birds can be defined as the fraction of nests that produce offspring. Research has been carried out into whether a correlation exists between breeding success and *cow density* on a meadow/the date from which the cows are grazing on the meadow (fig. 3).

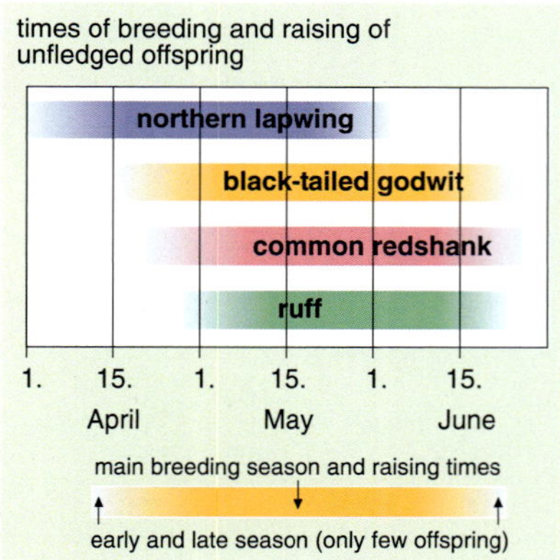


**2** Breeding period of wading birds

### Tasks

③ Analyse and compare the correlations between the breeding success of the individual wading bird species and the cow density on the meadow/date when grazing starts.

④ Give reasons why both the black-tailed godwit and ruff are highly endangered species.

⑤ Use the species mentioned to illustrate the conflict between agriculture and nature preservation. Suggest possible solutions.

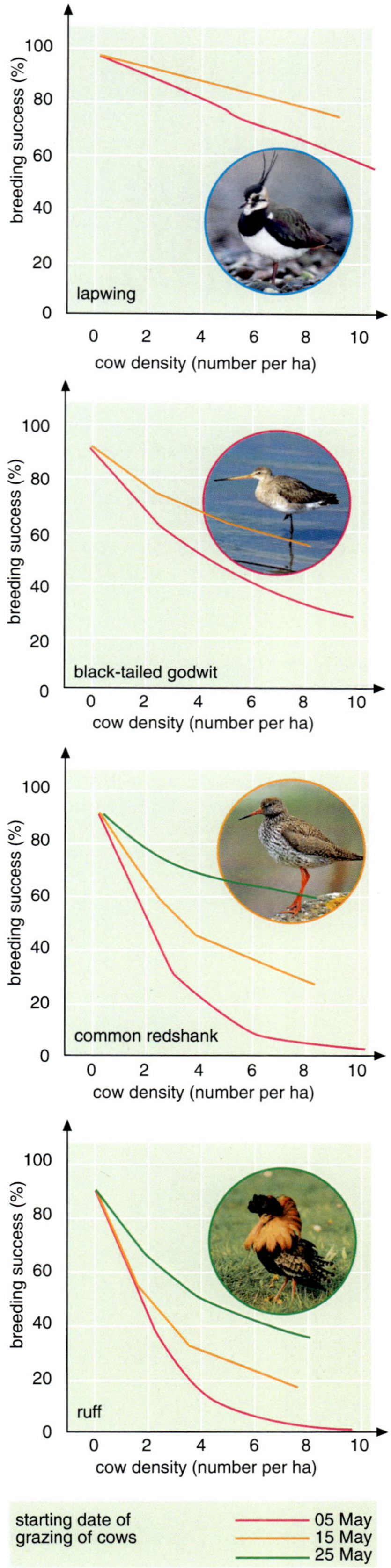


**3** Breeding success of wading birds

**Basic concepts**

# Regulation and control

Organisms and biocenoses keep many state variables within close limits, even when internal or external factors change for a short time (homeostasis). Through regulation systems, these state variables remain within a functionally adequate range. According to technical cybernetics, the aspired value is defined as the *set value.*
For certain procedures it is possible to adjust this value to changes that are happening. While a state can be maintained through regulation, control mechanisms can modify the direction or intensity of processes.

Negative feedback is a characteristic feature of regulation systems. Models of *control loop* systems in the form of block diagrams can show this by distinguishing between different components of a control loop. They are used to explain physiological processes and are usually complex. Arrow diagrams are easier to understand and are often used in population ecology.

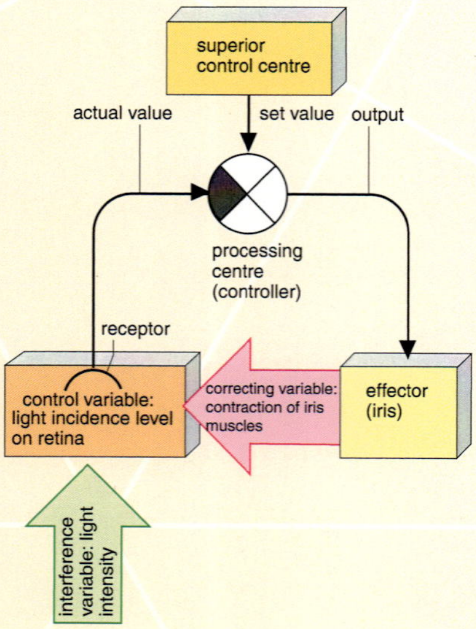


## Pupil reflex

The pupil reflex is an example for a proportional control system: with increasing light exposure the pupil becomes smaller, while less light exposure enlarges the pupil according to the current deviation from the set value.

Contrary to the simplest form of a control loop, in this system there is a superior control centre, which allows the adjustment of the set value. In simplified terms this can be compared with a water level control system.

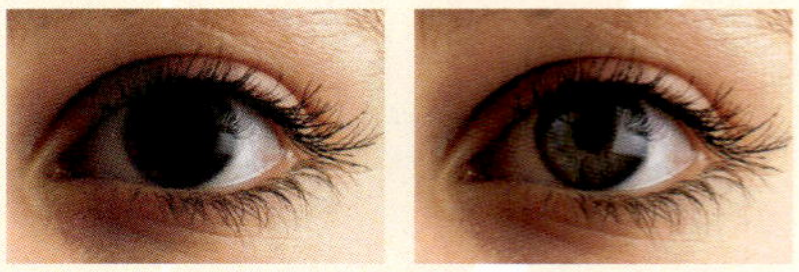

**Pupil reflex**

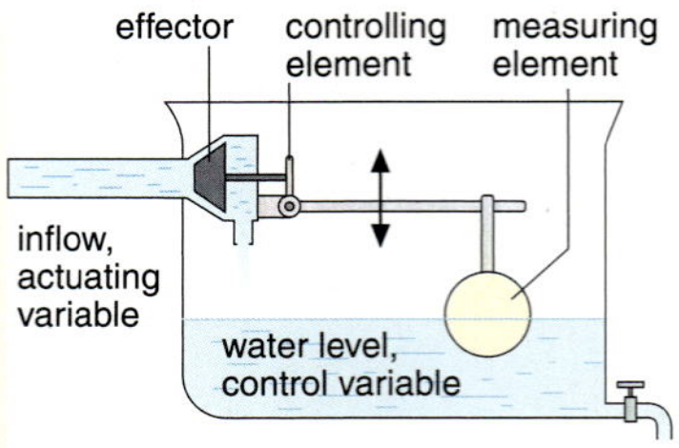


automatic water level regulation

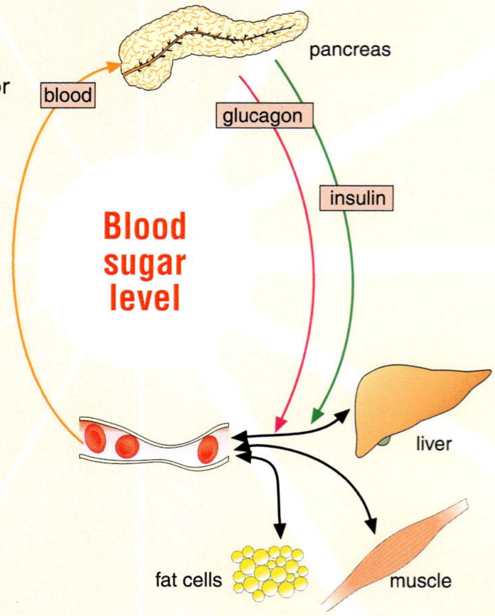


## Blood sugar level

The level of blood sugar is regulated by several interlocking control loops (networked systems). The concentration of blood sugar is constantly measured and maintained within a range of 60 to120 mg glucose per 100 ml of blood. Two hormones produced in the pancreas (insulin, glucagon) are involved in this regulation system. Their activity is influenced by other organs, nerve signals and other hormones. Factors like stress, muscle activity or times of rest act as disturbances.

## Body temperature

Birds and mammals are able to maintain an almost constant body temperature. Two very different effectors are involved in this regulation system: muscle contractions are used to actively release heat; sweating and panting is used to cool down.
For poikilothermic animals, behavioural changes can sometimes act as effectors: reptiles living in high altitudes or in the steppe regulate their body temperature through sunbathing during the day and by retreating to protected caves during the night.

**Body temperature**

● Think about connections ● See connections ● Develop connections

## Examples from technology

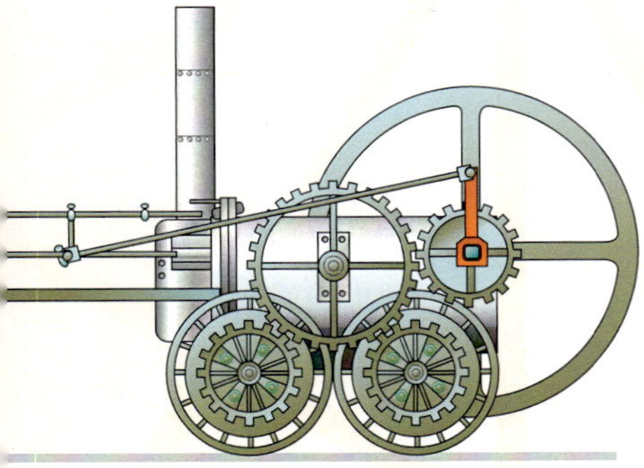

### Examples from technology

Control loop models mirror models used in technology: thermostats of heaters and refrigerators or the controls of an iron keep *preset* temperatures constant. However, since the construction of the first steam engine, other regulatory principles have also played a role in industry and technology: in 1788 JAMES WATT (1736 – 1819) developed a *flyweight governor* that ensured a constant output performance of steam-powered machines. Control systems for coolants in cars are e.g. linked with pipe systems that have the maximum area for heat exchange possible.

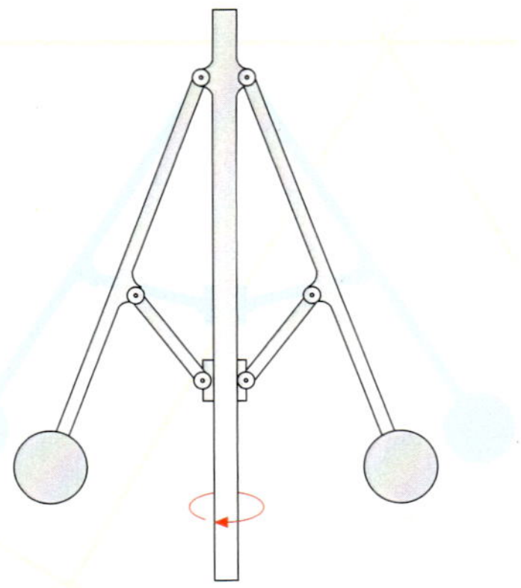

### Population size

The population sizes of different species of a biocenosis are interdependent. Feedback systems deliver only temporarily stable conditions; they can change quickly to reflect new environmental conditions. A defined set value does not exist.

An increase in the number of individuals can for example intensify intraspecific competition; it can lead to an oversupply of resources for predators or parasites and to an increased danger of spreading of infectious diseases.

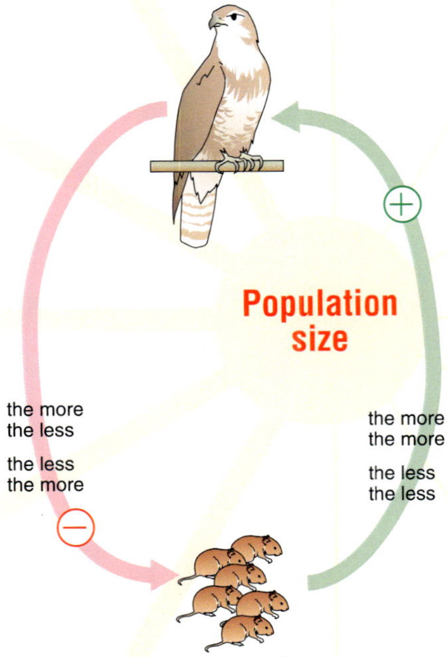


### Genetic and enzymatic activity

Repression of end products and substrate induction are examples for a negative gene regulation as they are controlled by the feedback of repressor proteins. Positive gene regulation is arranged through the attachment of activating proteins on a control region. In both processes allosteric proteins play a role. They can modify their spatial structure through the attachment of effectors *(conformational change)*.

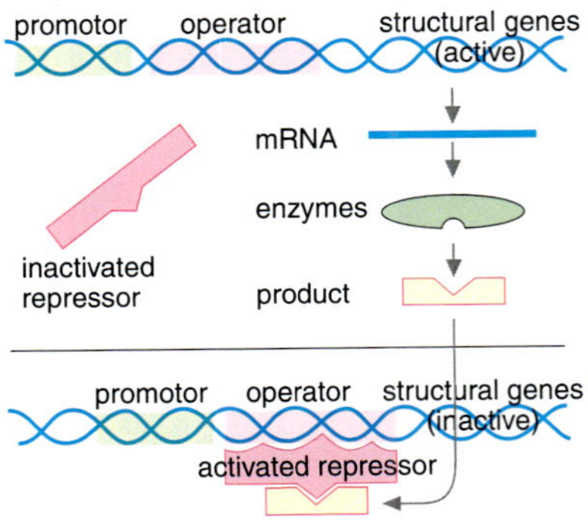


## Genetic and enzymatic activity

# Tasks

① Clarify the differences between regulation and control systems using the examples of motorised toy boats with and without remote control.

② Contrary to the pupil reflex, the regulation of blood pressure is an example of 'temporal integral regulation': despite the influence of permanent disturbance variables, the regulation process achieves and maintains a must value. Explain this with regard to blood loss due to injury.

③ The principle of negative feedback reduces or completely compensates for deviations from a set value through counteractions. Other models are mechanical equilibria, overflow systems, chemical buffering or 'feed-forward systems'. Find information about these models and comment on differences.

④ Positive feedback is self-intensifying. Find an example for this statement in the field of both biology and sports. Explain how each example works.

⑤ Muscles receive signals to contract both via $\alpha$ motor neurons and via $\beta$ fibres, which lead to muscle spindles. Prepare a short presentation about the mutual influences between muscles and muscle spindles with regard to contraction and elongation.

**Basic concepts**

# Genetic variation and evolutionary adaptation

The limbs and body structures of vertebrates often give indications about their habitat and way of life. Such a relation between structure, function and environment is true for all living beings. When a structure and the associated function supports and promotes the survival of an organism, this is defined as *evolutionary adaptation*. Hares are a good example; they are well camouflaged by their fur when they push themselves towards the ground. This evolutionary adaptation, a gradual genetic change that happens over many generations, is often distinguished from the process of adaptation, which signifies individual adaptations as displayed by the chameleon or which can be caused by modifications.

Evolutionary adaptation is only possible because of *genetic variation* and is caused by natural selection. Variation means that there are slight differences between individuals, cells, programs, functions, structures and strategies. Some of these variants can meet the demands of living conditions better than others. If this phenomenon is genetically determined, the variants are subject to natural selection and either have an advantage for survival or they become extinct. Over many generations, this results in the evolutionary adaptation to the current environmental conditions.

### Plasticity

The ability of living beings to individually adapt is based on *genotypes* which, depending on the environment, can be expressed in different *phenotypes* (contrary to genetically determined *polymorphism*); this results in visible *modifications*. The nervous system also adapts anatomically to experiences that have been made (neuroplasticity). From an ecological point of view, plasticity represents the ability of organisms to exist within certain areas of one or more environmental factors for a long period of time (reaction curve).

## Shape and environment

### Polymorphism

Polymorphism means that, due to genetic factors, individuals of a population have different appearance but are not defined as subspecies or races. There do not have to be any continuous transitions between the different phenotypes, as the example of the dimorphism of the *map butterfly* shows. Some white clover plants release *hydrocyanic acid* in response to herbivores; however, they do not protect themselves if only a small number of snails is present. Plants releasing hydrocyanic acid are more sensitive towards frost and infections.

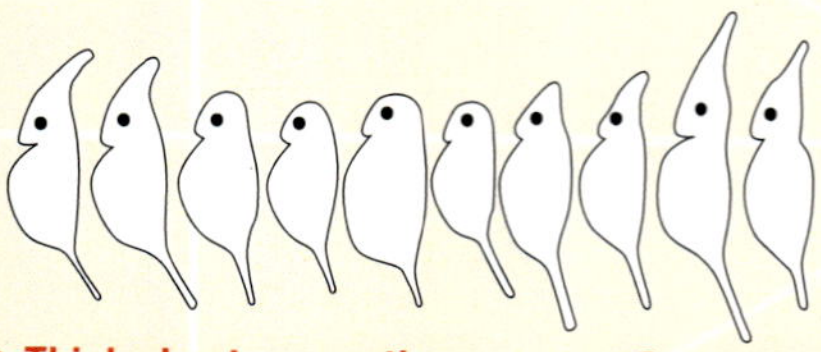

## Plasticity

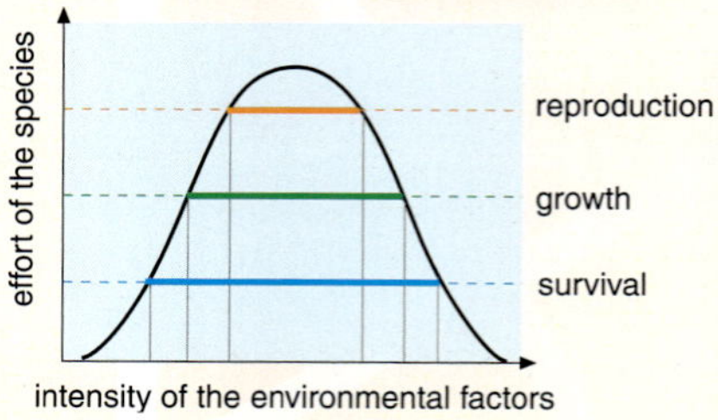


### Shape ("gestalt') and environment

Independent of the relation of different species, but due to analogous ways of life, similarities can develop with regard to body shape or certain organs. A well known example for this *convergence* is the streamlined body shape of different vertebrate species. Conversely, environmental conditions permitting, founding populations can spread into diverse new species, as the example of Darwin's finches shows *(adaptive radiation)*. Frogs or insects change during the course of their individual development *(metamorphosis)*.

## Polymorphism

● Think about connections ● See connections ● Develop connections

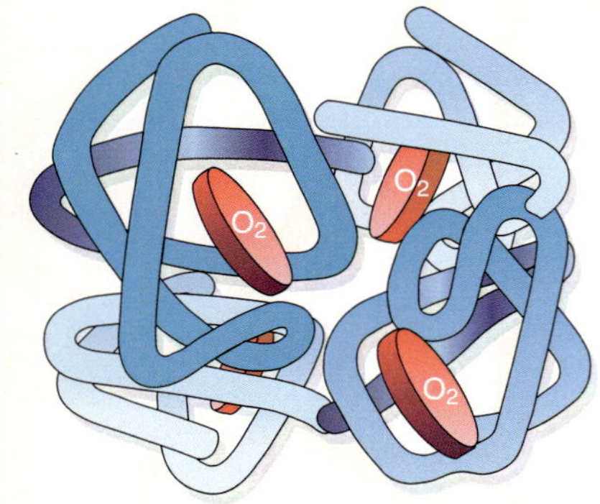


## Biochemical adaptation

### Biochemical adaptation

Organisms are able to adjust most metabolic processes in response to changing environmental conditions and manage to maintain the adjusted values through regulation thereafter. Humans living at high altitudes, for example, are adapted to lower oxygen supply by having *higher concentrations of haemoglobin* in their blood. At very low temperatures, insects and amphibians produce *antifreeze proteins* which help to avoid the formation of ice crystals in their body fluid. In the brain of rainbow trouts, acetyl cholinesterase varies depending on temperature. Increased muscle growth due to regular use can also be traced back to biochemical processes.

## Clone selection in the immune system

### Clone selection in the immune system

The immune system produces an immense amount of B cells, each of which can produce a different antibody. Only a cell which has identified and antigen will reproduce as a plasma cell and produce the antigen-fitting antibodies in large quantities. Thus only those antibodies are selected for production, which help to combat invading pathogens.

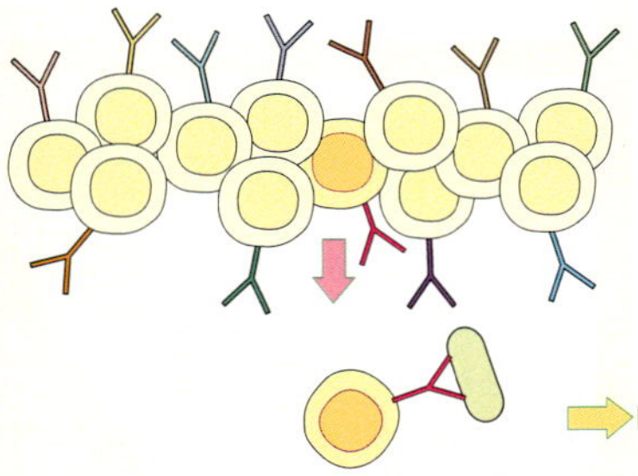

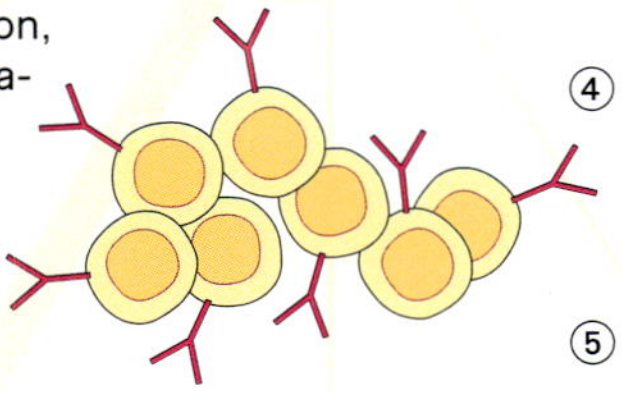

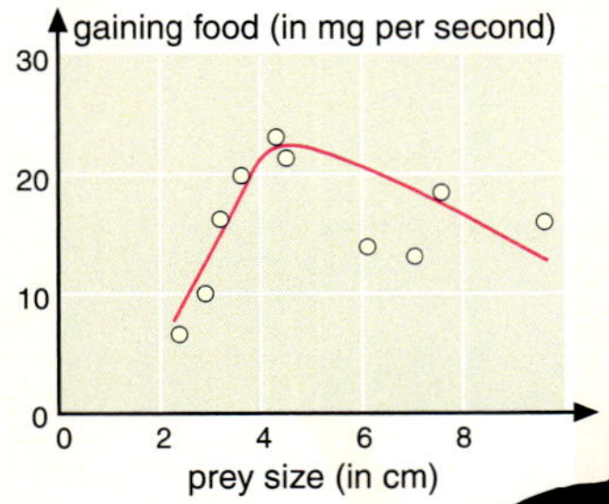


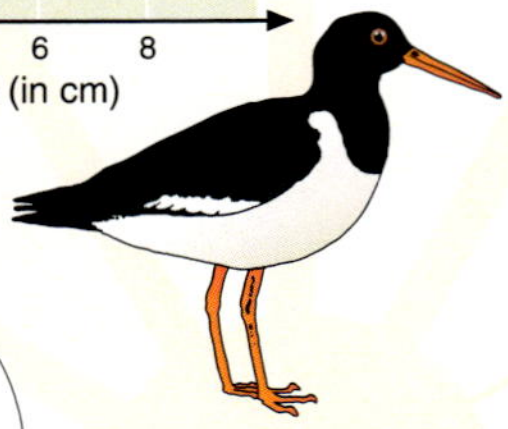

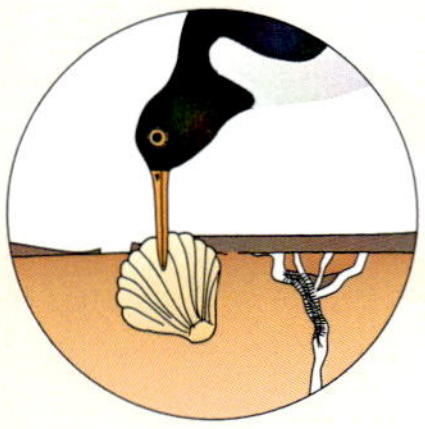

## Optimisation

### Optimisation

It can be the aim of optimisation to achieve maximum performance at a minimum cost and, from an economical point of view, to conserve resources. Like economists, biologists analyse costs and benefits, but in relation to evolutionary adaptations. Energy invested in a process counts as cost; benefits can for example relate to behaviour that maximises fitness *(adaptation value)*.

# Tasks

1. Selection is said to work according to the "lead shot principle". Explain what is meant.
2. Find one example each for genetically fixed polymorphism, balanced polymorphism and plasticity. Compare these examples.
3. Biochemical adaptations can be based on modifications in gene expression or changes in enzymatic activity. Explain.
4. Both high biodiversity and the fact that all organisms have common life processes can be explained by the theory of evolution. Give reasons for this statement.
5. Check whether the common phrase "survival of the fittest" is an adequate description for natural selection.
6. Nowadays many pathogens can no longer be treated with antibiotics. Describe how resistance against antibiotics occurs and show how this relates to genetic variation and evolutionary adaptation.
7. In industrial and medical research a certain substance is minimally modified, both deliberately and sometimes also by accident; then its effectiveness is tested out. Compare this process to the biological concept of evolution.
8. Compare the following examples with regard to adaptation and evolutionary adaptation: the recolouring of plaice, a stoat *(Mustela)* during winter and the camouflage of hares.

# Basic concepts

# History and relationship

"Nothing in biology makes sense except in the light of evolution". The statement by THEODOSIUS DOBZHANSKY (1900 – 1975) makes it clear that all biological phenomena can eventually be explained by evolutionary processes. All species existing today, including human beings, have a common origin. Seen over time, only the duration of the common development is different. The measure for the length of common history is the degree of relationship.

The diversity of life which can be observed today can be explained by history and relationship, which in turn are closely linked to the concepts of variation, evolutionary adaptation and reproduction.

### Life processes

Metabolism, cellular organisation, growth, reproduction, sensitivity, regulation, movement and evolutionary adaptations are characteristic features of any form of life. If all of these features apply, they define a living being. Since the emergence of organisms, these features have remained unchanged. Based on this, a high diversity of life has developed on all organisational levels.

**Life processes**

**Randomness**

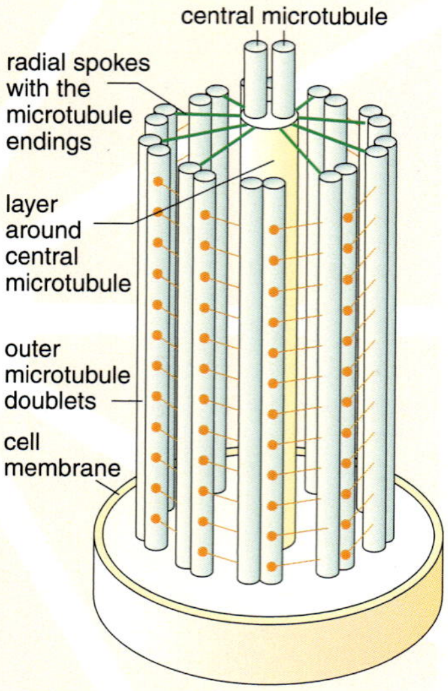


### Randomness

Mutations happen equally randomly as the fact that we have developed just five fingers or seven *cervical vertebrae*. However, some of these random events differ: for some objectively random developments there are no reasons to be found, while for other, subjectively random events, no explanations can be found in terms of evolutionary adaptations. The number of microtubules in flagella and cilia being in a 9 x 2 + 2 arrangement seems to have happened by chance; its common presence in all eukaryotes is explained by the ancestry of a common archetype.

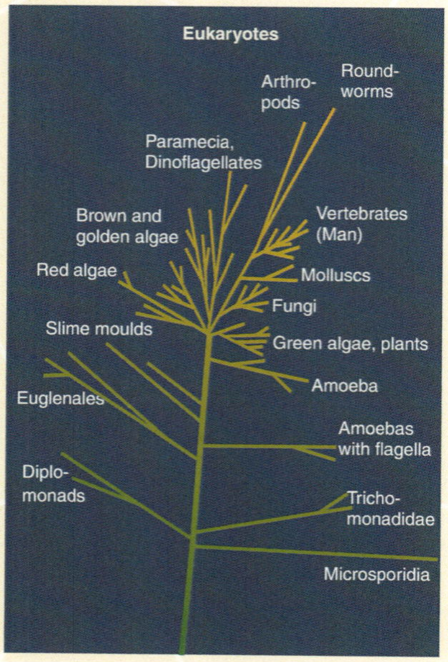


### Systematics

Relationships can be displayed, for example, in the form of phylogenetic trees. The phylogenetic tree has to be adjusted for each new finding. Archaea found in hot springs or black smokers (these are small volcanoes at deep-sea level) currently become more and more the focus of scientific research. These extremophiles prefer high temperatures or high sulphur concentrations, crude oil or salt lakes. Although the inner structure of archaebacteria places them close to bacteria, their biochemistry shows so many particularities that taxonomists place them in an exceptional position.

**Systematics**

● Think about connections ● See connections ● Develop connections

Uniqueness

### Uniqueness

Repeats of scientific experiments normally lead to the same results, provided they are conducted under the same experimental conditions. The course of evolution of life on Earth however is a historically unique process, which even under the same starting conditions could not happen in the same way again. As it can neither be repeated nor analysed with established scientific experimental methods, only indirect deductions, e.g. from *fossils*, are possible. Although species and individuals are unique, they share similar principles regarding structure and function.

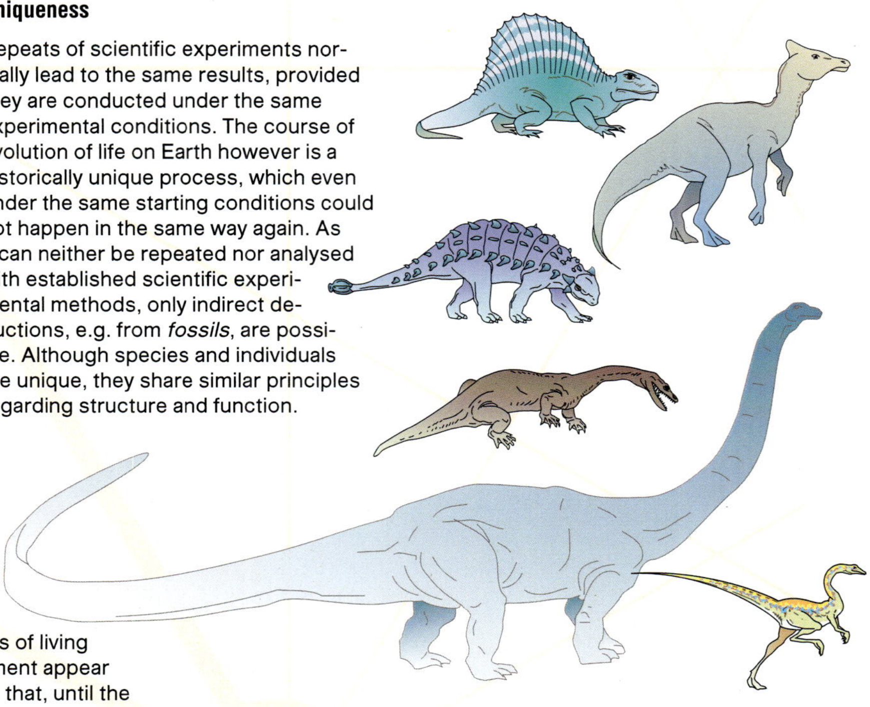

### Questions of function

Evolutionary adaptations of living beings to their environment appear so purposefully created that, until the principle of natural selection was formulated, they were regarded proof for the existence of God. Whereas other scientific disciplines research the questions of "What?", "How?" and "Why?", questions for which there are only descriptive and explanatory answers, biology in addition researches the question of function and biological significance: "What for?". The traits of living beings cannot be understood without an answer to this question. In all other natural sciences this question is not relevant.

Questions of function

Inter-actions

### Interactions

Flowers and their pollinators show that the structure of a flower can be exactly adapted to the animal species transporting pollen. Such mutual interactions lead to specialisations on both sides. This *co-evolution* is true for symbioses, predator-prey systems or parasite-host relationships. When foreign species enter an ecosystem, this process of mutual adaptation is missing. Extinction or mass proliferation of the foreign species can occur. If single groups, such as island populations, become isolated from this process, new species can evolve.

# Tasks

① Is co-evolution based on randomness? Discuss.
② Find examples for "purpose-built" things in biology, medicine and technology. What are the criteria of your choice?
③ Analyse the particular influences impacting on island populations, both from genetic and from evolutionary points of views.
④ C. Lyell (1797 – 1875) formulated the principle of uniformitarianism. He tried to explain the processes that happened in the past using only causal factors, which are still present today and can therefore be analysed. According to this, which evolutionary factors could have led to biological diversity?

hink about connections • See connections • Develop connections

## Glossary

**abiotic** [ˌeɪbaɪ'ɒtɪk]
Caused by the non-living environment (see *environmental factor)*

**Allen's rule** ['ælənz ˌru:l]
Distant body parts of warm-blooded animals are often smaller if a species populates cooler regions; close relatives populating warmer regions have larger body parts

**Autecology** [ˌɔ:ti:'kɒlədʒɪ]
Subfield of ecology; deals with the relationships of a species and its environment (see also *ecological factor, resource*)

**Bergmann's rule** ['bɜ:gmənz ˌru:l]
Close relatives of homoiothermic specimen often show a gradual increase in size from cooler to warmer climatic regions

**Biocenosis** [ˌbaɪəʊsɪ'nəʊsɪs]
Community of all organisms in a living space

**Biomass** ['baɪəʊmæs]
Total amount of mass (anhydrous mass) of all individuals in an environment; often referred to for a particular part of an ecosystem (e.g. populations, trophic levels, etc.)

**Biosphere** ['baɪəʊsfɪə]
The sum of all living spaces on Earth

**biotic** [baɪ'ɒtɪk]
Caused or influenced by living organisms (see *environmental factor*)

**Biotope** ['baɪəʊtəʊp]
Living space of a life community

**Consumer** [kən'sju:mə]
An organism using already existing biomass – from living organisms – as a resource (compare to *producer*)

**Control mechanism**
[kən'trəʊl ˌmekənɪzm]
Cybernetic model representing *control* in block diagrams or arrow diagrams

**Control mode** [kən'trəʊl ˌməʊd]
Unlike regulation the impact on direction or intensity of parameters or processes

**Cycle of matter** [ˌsaɪkl‿əv 'mætə]
Circulation of matter and chemical elements (such as carbon or nitrogen) in an ecosystem; a cycle consists of reservoirs (e. g. biomass, atmosphere, oceans, rocks) and flows (e. g. assimilation, decomposition, fossilisation); producers bind elements from the reservoirs atmosphere and soil in their biomass; consumers and decomposers usually release these elements back into the same reservoirs

**Decomposer** [ˌdi:kəm'pəʊzə]
An organism that feeds on dead organic matter and decomposes it; eventually only inorganic substances are left

**Detritus** [dɪ'traɪtəs]
Organic residual materials that are produced by the decay of organisms; these substances are deposited e. g. at the bottom of a body of water

**Diversity** [daɪ'vɜ:sətɪ]
Variety in an ecosystem; considers the numbers of species, their frequency distribution, and their genetic diversity. A common measurement is the diversity index according to SHANNON

**Ecoethology** [ˌi:kəʊi:'θɒlədʒɪ]
Scientific approach dealing with the adaptation of behavioural patterns – not only of social behaviour – and its interaction with ecological parameters

**Ecological factor** [ˌi:kəˌlɒdʒɪkl 'fæktə]
Factor that has an impact on population growth; a distinction is made between density-dependent factors, which are affected by the number of individuals present (e.g. competition for resources), and generally acting density-independent factors (e. g. weather, catastrophes)

**Ecological pyramid**
[ˌi:kəˌlɒdʒɪkl 'pɪrəmɪd]
Illustration of the trophic levels of an ecosystem; on lower trophic levels more organisms are often existent, providing higher levels with food; the succession can be presented as a pyramid

**Ecosystem** ['i:kəʊˌsɪstəm]
Dynamic network consisting of *biocenosis* (community of living beings) and *biotope* (living space) that is based on cycles of matter and energy flow

**Emission** [ɪ'mɪʃn]
Production of harmful substances which are spread over the surrounding area

**Energy flow** ['enədʒɪ ˌfləʊ]
Transfer of energy-rich substances within a food chain

**Environmental factor**
[ɪnˌvaɪərənˌmentl 'fæktə]
External impact on a living organism; *abiotic factors* of the non-living environment (physical, chemical) and *biotic factors* based on other living beings (e. g. prey, predator, competitor) have to be distinguished

**eurypotent** [ˌjʊərɪ'pəʊtənt]
Ability of an organism to tolerate strong fluctuations of a certain environmental factor

**Eutrophication** [ˌju:trɒfɪ'keɪʃn]
Accumulation of minerals within a body of water ("over-fertilisation")

**Feedback** ['fi:dbæk]
The return of the output of a state or process as input; positive influence (stimulatory effect) or negative influence (control mechanism)

**Food chain** ['fu:d‿tʃeɪn]
Succession of organisms feeding on each other, passing on biomass and energy to the next link; in many cases the chain starts with plants as *producers* and continues with *consumers* and *decomposers*

**Food strategy** ['fu:d ˌstrætədʒɪ]
Manner of choosing and acquiring food, which reflects requirements and anatomical abilities

**Food web** ['fu:d ˌweb]
Net-like combination of feeding relationships *(food chains)* within an ecosystem

**Fundamental niche** [ˌfʌndə'mentl ˌni:ʃ]
Summarises the requirements of a species under experimental conditions without the influence of biotic factors such as competition or predation (see *realised niche, ecological niche*)

**Fungicide** ['fʌndʒɪsaɪd]
Inhibiting the growth of fungi (see *pesticide*)

**Generalist** ['dʒenrəlɪst]
Living being that is able to exist under highly varying living conditions; a food generalist e.g. shows a broad diet spectrum (compare to specialist)

**Gross primary production**
[ˌgrəʊs ˌpraɪmərɪ prəˈdʌkʃn]
Total amount of biomass produced through photosynthesis or chemosynthesis over a certain period of time (see net primary production)

**Habitat** [ˈhæbɪtæt]
Preferred living space of a single species

**Herbicide** [ˈhɜːbɪsaɪd]
Substance damaging plants; used to extinguish weeds (see pesticide)

**homoeothermic** (warm-blooded)
[ˌhəʊmɪəʊˈθɜːmɪk (ˌwɔːmˈblʌdɪd)]
Organisms regulating their body temperature within given limits at a physiological optimum, more or less independent of the temperature of the surroundings; mammals and birds are homoeothermic

**Immission** [ɪˈmɪʃn]
Harmful substance in the air (*emission*) which is able to impact on living organisms and buildings

**Indicator organism**
(indicator species)
[ˈɪndɪkeɪtə‿ɔːgənɪzm (ˈɪndɪkeɪtə‿spiːʃiːz)]
A living being that is sensitive (stenopotent) to certain environmental factors; its presence or absence can be used as an indication about these environmental factors in an area; indicator organisms play an important role in evaluating and recording the pollution of water, air and soil

**K-strategist** [ˈkeɪ ˌstrætədʒɪst]
Living being that invests its energy mainly in ensuring its survival and is therefore very competitive and long-lived (compare to r-strategist)

**Landscape conservation area**
[ˌlændskeɪp ˌkɒnsəˈveɪʃn ˌeərɪə]
Area with special usage limitations; they mainly relate to building bans and to restrictions determining the change of a landscape. Usually, meadows may not be modified into cropland (see *nature protection area*)

**Liebig's law of the minimum**
[ˌliːbɪgz ˌlɔː‿əv ðə ˈmɪnɪməm]
The growth of plants is determined and limited by that mineral, which is in lowest supply compared to the demand situation

**Lotka-Volterra equations**
[ˌlɒtkə vɒlˈteərə‿ɪˌkweɪʒnz]
From the Lotka-Volterra model, three rules can be derived, describing the interaction of predator and prey populations and its reciprocal impact on the density of individuals;
1. Recurrent fluctuations of population densities
2. Long-term stability of the midpoints
3. Faster recovering of the prey population after having been depleted

**Nature protection area**
[ˌneɪtʃə prəˈtekʃn ˌeərɪə]
Area that is subject to special usage limitations based on a nature protection regulation; e. g. these can be limitations of recreational and agricultural usage (see also *landscape conservation area*)

**Net primary production**
[ˌnet ˌpraɪmərɪ prəˈdʌkʃn]
Amount of biomass produced through photosynthesis or chemosynthesis over a certain period of time after subtraction of the portion of energy used by the producers themselves (see gross primary production)

**Niche, ecological** [ˌiːkəˈlɒdʒɪkl ˌniːʃ]
Total of requirements made by the species on its environment; the demands are determined by the impact of environmental factors and by the usage of vital resources; ecological niche can also be described as the totality of ecological potencies

**oligotrophic** [ˌɒlɪgəʊˈtrɒfɪk]
Poor in minerals (most notably phosphate); often also means high oxygen content in standing body of water; opposite of eutrophic: high mineral content (see *eutrophication*)

**Parental investment**
[pəˌrentl ɪnˈvestmənt]
Parental care invests in offspring and therefore raises the probability of its survival; however, to some extent this behaviour impedes parents from producing more offspring

**Pesticide** [ˈpestɪsaɪd]
Toxin combating undesirable insects *(insecticide)*, fungi *(fungicide)* or plants *(herbicide)*; predominantly applied in agriculture, forestry and gardening

**Phytoplankton** [ˌfaɪtəʊˈplæŋktən]
Herbal microorganisms in bodies of water (see *plankton*)

**Plankton** [ˈplæŋktən]
Mostly small aquatic organisms that can only to a minor degree move actively; predominantly moved by currents; a distinction is made between herbal plankton (phytoplankton) and animal plankton (zooplankton)

**poikilothermic** (cold-blooded)
[ˌpɔɪkɪləʊˈθɜːmɪk (ˌkəʊldˈblʌdɪd)]
Body temperature of organisms is dependent on the surrounding temperature

**Population** [ˌpɒpjəˈleɪʃn]
Group of conspecific individuals within a certain region representing a reproductive community

**Potency, ecological**
[ˌiːkəˌlɒdʒɪkl ˈpəʊtənsɪ]
Extent of response of a species to an environmental factor; quantified as vitality (ability to survive; e. g. biomass increase, survival rate) and described in relation to this environmental factor

**Primary production**
[ˌpraɪmərɪ prəˈdʌkʃn]
Organic biomass made by producers at the beginning of a food chain and at the bottom of an ecological pyramid; refers to a certain period of time; a distinction is made between the biomass produced overall (gross primary production) and the amount of biomass remaining after subtraction of own consumption by the producers

**Producer** [prəˈdjuːsə]
Organism producing biomass through photosynthesis or chemosynthesis; autotrophic plants represent the largest group (compare to consumer)

**Realised niche** [ˈrɪəlaɪzd ˌniːʃ]
Requirements of a species under outdoor conditions (compare to *fundamental niche*; *niche, ecological*)

**Regulation** [ˌregjəˈleɪʃn]
Compensation for disturbances and the achievement of a state of equilibrium through negative feedback mechanisms

**Reproductive strategy** [ˌriːprəˌdʌktɪv 'strætədʒɪ]
Considers the portion of available biomass and energy a species invests in reproduction; the portion is genetically determined

**Resource** [rɪ'zɔːs]
Constituent parts of the environment that are needed and are used up by an organisms and are therefore not available to any other organism (food, water, space); shortage of resources results in competition

**r-strategist** ['ɑː ˌstrætədʒɪst]
Living being that mainly invests its energy in reproduction; lots of offspring, but not competitive

**Saprobic system** ['sæprəbɪk ˌsɪstəm]
Indication system for water pollution based on the presence or absence of indicator organisms (see *indicator organisms*)

**Secondary production** [ˌsekəndrɪ prə'dʌkʃn]
Biomass produced by consumers based on assimilated food

**Specialist** ['speʃəlɪst]
Living being with limited potency; very well adapted to stable environmental conditions (see also *generalist)*

**stenopotent** [ˌstenəʊ'pəʊtənt]
Feature of living being that can only survive in a narrow range of an environmental factor (compare to *eurypotent)*; little tolerance concerning an environmental factor

**Succession, ecological** [ˌiːkəˌlɒdʒɪkl sək'seʃn]
Natural progression of ecosystems, e.g. reforestation of fallow land or aggradation of a body of water

**Synecology** [ˌsɪniː'kɒlədʒɪ]
Subdiscipline of ecology; deals with the relationships of living beings in a biocenosis (life community, e. g. predator-prey relationships, parasitism, etc.)

**System** ['sɪstəm]
Unit with defined limits consisting of parts (elements) and their interactions (relations)

**Territory** ['terɪtərɪ]
Living beings defend their living space against conspecifics, ensuring provision of resources

**Tolerance region** ['tɒlərəns ˌriːdʒən]
Range of a certain environmental factor, in which a living being is able to exist

**Trophic level** ['trɒfɪk ˌlevl]
1. Degree of mineral supply in bodies of water and soils: oligotrophic, mesotrophic, polytrophic or eutrophic (see *eutrophication*);
2. Position of living beings within the *food web* (the *food chain*): *primary producers* and various *consumers*

**van't Hoff's rule** [ˌvænt'hɒfs ˌruːl]
Dependence of the rate of a chemical reaction on the surrounding temperature; generally the speed of a reaction is increased twofold or threefold for each rise of 10 °C

**Zooplankton** [ˌzuːəʊ'plæŋktən]
Animal microorganisms within a body of water; drifting in water and predominantly moved by currents (see *plankton*)

# Pictures sources

6.1 Getty Images (Stone/Laurie Campbell), München – 6.2 Getty Images (Taxi/Neil McIntyre), München – 6.3 Getty Images (Stone/ Bruno Dittrich), München – 6.4 Corbis (Jim Craigmyle), Düsseldorf – 6.5 Angermayer (Ernst Elfner), Holzkirchen – 7.1 CSIRO Publishing, Collingwood – 7.2 Okapia (Manfred Danegger), Frankfurt – 7.3 Reinhard-Tierfoto, Heiligkreuzsteinach – 7.4 Corbis (Joe Bator), Düsseldorf – 8.1 Angermayer (Hans Reinhard), Holzkirchen – 8.2 Okapia (Barbara Rios/NAS), Frankfurt – 11.S Getty Images, München – 11.1 Okapia (Joe McDonald), Frankfurt – 12.S Stockbyte, Tralee, County Kerry – 14.1 Dr. Rudolf König, Preetz – 14.Rd. FOCUS (SPL), Hamburg – 15. Rd. Dr. Heiko Bellmann, Lonsee – 16.1 Okapia (Erich Thielscher), Frankfurt – 17.1 Reinhard-Tierfoto, Heiligkreuzsteinach – 18.S Prof. Dr. Ulrich Heitkamp, Gleichen-Diemarden – 19.S Silvestris (Walter Rohdich), Dießen – 21.1 Angermayer (Hans Pfletschinger), Holzkirchen – 22.Rd. Reinhard-Tierfoto, Heiligkreuzsteinach – 24.1 Okapia (Robert Groß), Frankfurt – 24.2a Angermayer (Hans Reinhard), Holzkirchen – 24.2b Okapia (Luiz C. Marigo), Frankfurt – 24.2c FOCUS (Andrew Syred, Science Photo Library), Hamburg – 24.2d Okapia (Stouffer), Frankfurt – 24.2e Okapia (Robert Henno, Wildlife Pict.), Frankfurt – 24.2f Klett-Archiv, Stuttgart – 25.1 aus „David Attenborough, The atlas of the living world“, Houghton Mifflin Company, Boston 1989, S. 124/125 (Giani Tortoli) – 25.Rd. Angermayer (Hans Reinhard), Holzkirchen – 26.1 Okapia (Stouffer), Frankfurt – 28.1 Corbis (Steve Terrill), Düsseldorf – 28.2 Creativ Collection Verlag GmbH, Freiburg – 29.1 Okapia (Sven Halling/Naturbild), Frankfurt – 29.2 Angermayer (Hans Reinhard), Holzkirchen – 30.1, 31.1 Naturfotografie Frank Hecker (F. Sauer), Panten-Hammer – 32.Rd. Okapia (Cattlin/Holt Studios), Frankfurt – 33.Rd. Angermayer (Hans Reinhard), Holzkirchen – 37.2 Okapia (Hans Reinhard), Frankfurt – 37.Rd.1-3 Dieter Schmidtke, Schorndorf – 40.S Fotosearch RF, Waukesha, WI – 40.1 Angermayer (Hans Pfletschinger), Holzkirchen – 40.2 Reinhard-Tierfoto, Heiligkreuzsteinach – 40.3 ZEFA (M. Mehlig), Düsseldorf – 40.4 Okapia (Cyril Ruoso/BIOS), Frankfurt – 40.5 Mauritius (Nakamura), Stuttgart – 40.6 IFA (BCI/Bob & Clara Calboom), Ottobrunn – 40.7 Bernhard Knauer, Göttingen – 40.8 Mauritius (Hänel), Stuttgart – 41.1-7 Okapia (Clyde H.Smith), Frankfurt – 41.8 Okapia (Wolfgang Lummer), Frankfurt – 42.1 Corbis (Yann Arthus-Bertrand), Düsseldorf – 44.1a/b Bernhard Knauer, Göttingen – 46.1a Corbis (RF Royalty Free), Düssel-dorf – 46.1b Corbis (Francesc Muntada), Düsseldorf – 48.S Tilman Wischuf, Cleebronn – 52.Rd. Bärbel Knabe, Leipzig – 53.S Getty Images (PhotoDisc), Hamburg – 54.3 Bildarchiv Sammer, Neuenkirchen – 56.2 Cartoon-Caricature-Contor (Mandzel), Pfaffenhofen – 58.S Superbild (Eric Bach), Unterhaching – 58.1, 3 Astrofoto (NASA), Sörth – 58.2 Silvestris, Dießen – 60.1 Schlosser, Fluterschen – 60.Rd. Okapia (Kjell-Arne Larsson), Frankfurt – 61.1, 2 aus „Die Düngung von Waldbäumen“ (Baule, Fricker), BLV, München – 61.3, 4 ZELMI (Prof. Hafner), Berlin – 61.5, 6 Prof. Dr. Ingrid Kottke, Tübingen – 61.7-9 Horst Hooge, Gleichen-Benniehausen – 62.1a Okapia (Falck), Frankfurt – 62.1b Helga Lade (D. Rose), Frankfurt – 62.Rd. Okapia (Hans Reinhard), Frankfurt – 64.1a Okapia (Detlef Lämmel), Frankfurt – 64.1b Okapia (Skip Moody), Frankfurt – 64.1c Okapia (Eduard Schäfer), Frankfurt – 64.1d Okapia (Winfried Wisniewski), Frankfurt – 64.1e Okapia (Dr. Peter Wernicke), Frankfurt – 64.2 Okapia (Jeffrey Rich), Frankfurt – 64.Rd.1 Okapia (L. Martinez), Frankfurt – 380.2 Okapia (Hans Reinhard), Frankfurt – 380.1a Okapia (H. Farkaschovsky), Frankfurt – 64.Rd.2 Okapia (Fritz Pölking), Frankfurt – 64.Rd.3 Okapia (Johan de Meester), Frankfurt – 65.S, 3a Corbis (Niali Benvie), Düsseldorf – 65.1 Okapia (G. Büttner/Naturbild), Frankfurt – 65.3b Okapia (Manfred Danegger), Frankfurt – 65.3c Okapia (Bernard Castelein), Frankfurt – 65.3d Angermayer (Günter Ziesler), Holzkirchen – 66.1a/b Okapia (Ulla Spiegel), Frankfurt – 66.2 Tilman Wischuf, Cleebronn – 68.1a Silvestris (J. & Ch. Sohns), Dießen – 68.1b Okapia (Konrad Wothe), Frankfurt – 68.2 Silvestris (Gerhard Kalden), Dießen – 68.3 IFA (The Natural History), Düsseldorf – 69.1 Corbis (Michael Kevin Daly), Düsseldorf – 70.1 IFA (Index Stock), Düsseldorf – 71.1 Okapia (Fritz Hanneforth), Frankfurt